WS

The Henry A. Wallace Series
on Agricultural History and Rural Life

IOWA STATE UNIVERSITY PRESS / AMES

WS

The Henry A. Wallace Series
on Agricultural History and Rural Life

R. Douglas Hurt, SERIES EDITOR

To Make a Spotless Orange

To Make a Spotless Orange

Biological Control in California

RICHARD C. SAWYER

The Henry A. Wallace Series
on Agricultural History and Rural Life

IOWA STATE UNIVERSITY PRESS / AMES

RICHARD C. SAWYER, a native of the citrus belt of southern California, received a B.S. in biology from the University of California, Riverside, and a Ph.D. in the history of science from the University of Wisconsin, Madison. Currently he teaches high school biology and mathematics.

© 1996 Iowa State University Press, Ames, Iowa 50014

All rights reserved

Authorization to photocopy items for internal or personal use, or the internal or personal use of specific clients, is granted by Iowa State University Press, provided that the base fee of $.10 per copy is paid directly to the Copyright Clearance Center, 27 Congress Street, Salem, MA 01970. For those organizations that have been granted a photocopy license by CCC, a separate system of payments has been arranged. The fee code for users of the Transactional Reporting Service is 0-8138-2755-8/96 $.10.

♾ Printed on acid-free paper in the United States of America

First edition, 1996

Library of Congress Cataloging-in-Publication Data

Sawyer, Richard C.
    To make a spotless orange: biological control in California/Richard C. Sawyer.
       p.   cm.—(The Henry A. Wallace series on agricultural history and rural life)
    Includes bibliographical references (p.   ) and index.
    ISBN 0-8138-2755-8
       1. Biological pest control agents—California—History. 2. Agricultural pests—Biological control—California—History. 3. Citrus fruits—Diseases and pests—Biological control—California—History. 4. University of California, Riverside. Citrus Research Center and Agricultural Experiment Station—History. I. Title. III. Series.
Sb975.5.U6S38   1996
634′.304996′09794—dc20   96-20194

Last digit is the print number: 9 8 7 6 5 4 3 2 1

TO JANET

Contents

Series Editor's Introduction

Today many people question the use of chemical pesticides for agricultural purposes and advocate the use of biological, "natural," methods to control pests that damage crops. The desire for scientists to discover biological control methods to combat insects harmful to crops is not a new idea. From the late nineteenth to the mid-twentieth century, a small group of scientists in California attempted to develop biological controls that would make the use of chemical pesticides unnecessary and at the same time help protect human health and the environment.

California provided a unique laboratory for agricultural scientists, with its particular crops, climate, and scientific organizational structure within the state government and university system. Moreover, California's citrus growers supported biological pest-control measures as the most practical, efficient, and cost-effective long-term methods to combat insects and produce cosmetically ideal, unblemished fruits that would provide irresistible appeal to consumers across the nation. The initial success of entomologists in fighting the cottony-cushion scale in the citrus orchards of southern California with the vedalia beetle, a "lady bug," which they imported in late nineteenth century from Australia, brought false hopes that a host of pests could be controlled, if not destroyed, by other farmer-friendly insects.

Despite repeated failures, entomologists conducted extensive research to find bugs that would fight other bugs for beneficial, that is, agricultural purposes. They gained financial support from the state government and citrus industry, even though most entomologists at the University of California and the state experiment station concentrated on the use of chemical controls. By the mid-twentieth century, high cosmetic standards, the large capitalization required for the maintenance of orchards, and the competitive nature of the citrus industry increasingly turned growers away from theoretical biological control techniques to practical chemical eradication methods. A host of new pesticides, particularly DDT, provided relatively cheap and effective chemical means to control insects, so the economic entomologists in the United States Department of Agriculture and in other states essentially ended biological control research. Some entomologists in California, however, continued their biological control investigations until the late 1960s, when a new approach to the control of insects, called integrated pest management, which is essentially the use of chemical, cultural, and biological methods, superseded their work.

Richard C. Sawyer has written a path-breaking history of the biological attempts to control insects in the fields, vineyards, and orchards of California. Specifically, Sawyer traces the work of the entomologists who attempted to provide scientific justification for the biological control of insects by introducing natural enemies, that is, other insects, to destroy the damaging pests that preyed on the citrus crop, particularly oranges. Sawyer also astutely analyzes the scientific, political, and institutional relationships that made California unique in the development of scientific and economic entomology. By so doing, he has made an important contribution to the history of agricultural science, the making of public policy, and the development of agribusiness.

R. DOUGLAS HURT

Acknowledgments

Much as I would like to take sole credit for this book, I could not have produced it without the advice and assistance of a great many others.

My professional interest in biological control began in the lab of S. D. Van Gundy and Diana Freckman at the University of California, Riverside, while I was an undergraduate. As I began to pursue the history of the subject, I received useful advice from John Cooper, William Coleman, Ronald Numbers, Eugene Cittadino, Ronald Tobey, Harry Lawton, Sharon Kingsland, Morton Rothstein, and Pete Daniel. Donna Hartley graciously shared her thoughts on a study she had once thought to do herself. Harry Coppel and Mallory Boush got me started learning some entomology. John Neu spotted background sources I would otherwise have missed. Gordon Gordh and Kenneth Hagen made contacts and found work space for me at Riverside and Berkeley, respectively. At the National Museum of American History, Mary Dyer made researchers feel like family.

The following helped me find or use manuscript sources: Cliff Wurfel, Pamela Hoatson, and Jack Hall at UC Riverside; William Roberts and Robin Rider at UC Berkeley; Roger Berry at UC Irvine; Paul Arnaud at the California Academy of Sciences; Richard Crawford and Charlie Roberts at the National Archives; Louie Hutchins, Pam Henson, and Bill Deiss at the Smithsonian Institution; Alan Fusonie and Judith Ho at the National Agricultural Library; Vince Moses at the Riverside Municipal Museum; Ron Baker at the Riverside Public

Library; Leo Bald at Sunkist Growers; Lloyd Andres at the USDA Albany lab; and Rosanne Clayton at the Australian Academy of Science. Theodore W. Fisher provided access to UCR Department of Entomology photographs.

I thank the following for their recollections and their patience with my questions: Daniel Aldrich, Al Boyce, Janet Mabry Boyce, Donald Clancy, Paul DeBach, Everett "Deke" Dietrick, Dick Doutt, Mack Dugger, Ted Fisher, Charles Fleschner, Ken Hagen, Irv Hall, Ivan Hinderaker, Carl Huffaker, Laurence Jones, Chuck Kennett, Earl Oatman, Ray Smith, Mabry Steinhaus, and Harvey Sweetman. Dick Smith, Larry Smith, Alice Smith, and Harriet Smith Hanson shared reminiscences of their father.

The following read at least one draft of the entire manuscript and gave valuable criticism: Ron Numbers, Tom Broman, Gregg Mitman, Allan Bogue, Vic Hilts, Robb Reavill, John Perkins, and an anonymous reviewer. Douglas Hurt, Linda Speth, and Jane Zaring have given great encouragement as they shepherded the book into print at Iowa State University Press.

Fellowships from the University of Wisconsin, Madison, and the Smithsonian Institution supported some of my research. Much of it would not have been possible without accommodations from members of my own family, particularly my parents, Ray and Darlene Sawyer; my sister, Kathleen Sawyer; and my grandmother, Dorothy Balogh. The National 4H Council helped make work in Washington affordable. Extra special thanks to Richard and Sandy Carlson, who made me a member of their family during a longer stay in southern California than they or I expected.

Faculty, students, and staff of the Departments of History of Science and History of Medicine at the University of Wisconsin, Madison, provided an encouraging, friendly atmosphere for scholarship. The staffs and facilities of the various campus libraries at Madison, Riverside, and the University of Pennsylvania helped keep the project moving. Thanks to Robert Auerbach and his lab for all the international food and those nighttime printing runs.

Finally, I cannot express how much I owe to the love and support of my wife, Janet Ewart, who has had to keep me steady while worrying about her own research.

Introduction

In 1962, marine biologist Rachel Carson ignited a fury of protest over the indiscriminate use of insecticides, DDT in particular. Not only were there technological disadvantages, she wrote in *Silent Spring,* but the chemicals seriously endangered human health and the environment. Pesticides poisoned unintended victims, such as birds and fish, and threatened the health of consumers and farm workers. Aimed especially at farm and forest pests, chemicals brought about their own obsolescence by producing resistant insects. One chemical replaced another as the pesticide industry tried to keep up with this phenomenon. Insecticides even created new pest problems by disrupting natural processes that had kept many kinds of insects from reaching damaging numbers.[1]

The drawbacks of chemical insecticides, Carson wrote, demanded another approach to pest control. The last problem, pest outbreaks actually caused by chemicals, pointed the way. Pesticides made the farmer's situation worse by killing natural enemies that had controlled potential pest species without human intervention. Deliberate use of such natural enemies, or *biological control,* was the alternative to chemicals. Carson defined biological control broadly. She included some methods proven effective only within the previous few years, and others that were just tantalizing hopes in her day and are little more than that now. For most of its history, however, biological control meant only other animals or diseases that killed the pests.

Permanent, harmless to humans, and safe for the environment, biological control was, in Carson's words, "the other road," the road we would have to take to avert ecological disaster. Reduction of pesticide use became a central objective of the environmentalist movement that she helped to launch. This goal would require a return to more natural means that American agriculture supposedly discarded long ago.

Biological methods had never disappeared entirely. Neither had they ever held the dominant position Carson wished for them. An entomological specialty of biological control developed over the seventy-five years before she published her controversial book. As a scientific enterprise, biological control secured a unique institutional and agricultural setting in California. In that setting, "the other road" survived the increasing attention paid to chemicals and was still in place when Carson wrote *Silent Spring*. Few states showed any serious interest in biological control after the first effective commercial pesticides appeared late in the nineteenth century. Not one other state followed California's example of employing scientists to search for natural enemies, although the Territory of Hawaii did.

This study follows the growth of the California program and its interaction with the agricultural and scientific communities, moving from a single, spectacular success that began the "bug vs. bug" era in the 1880s to the challenge of new pesticides and the broadening of natural enemy study on the eve of Rachel Carson's dire warning.

"Biological control" has three related, but distinct, meanings. First, it names an ecological phenomenon, the action of natural enemies to keep pest populations lower than they would otherwise be. Second, the phrase refers to the practice of deliberate introduction or manipulation of natural enemies to increase this effect. Third, the study of such naturally occurring or deliberately imposed action, together with the investigation of associated biological phenomena and the development of scientific theory to explain them, constitute a specialized discipline called biological control.[2]

Although a biological control project might involve no insects at all, as in the case of a virus used to control rabbits in Australia, the field remained primarily a part of entomology. The vast majority of biological control efforts were directed toward insect pests. Most of the rest used insects as biological agents to control weeds.

In the early days of biological control, some participants lacked any scientific background. People serving in political rather than scientific institutions nevertheless addressed themselves to entomologists. Those who were educated—formally or informally—as scientists counted themselves as entomologists and sought the approval of that discipline. Professional specialists in biological control emerged from the 1920s onwards, but always as entomologists first. Entomological literature remained the principal outlet for publication. Even in California, the only state to separate biological control institutionally, boundaries were drawn at a subdisciplinary level.

Although the phrase "biological control" first appeared in 1919, the idea emerged much earlier. After deliberately introduced natural enemies controlled a citrus pest known as cottony-cushion scale (*Icerya purchasi*) in California in 1889, entomologists recognized this method as a distinct form of pest control. The concept took various names, such as "the parasite method" and "bug vs. bug," before the current expression evolved. Writing on American applied entomology in 1913 after visiting the United States, German entomologist Karl Escherich termed the practice *biologische Bekampfung* and *die biologische Methode*. An American reviewer translated the latter as "the biological method of controlling insects." Leland O. Howard, longtime chief of the U.S. Bureau of Entomology, proposed "the biological method of fighting insects" as a standard label in 1916. Howard's devotee Harry S. Smith, the central figure of this study, altered the label to "the biological method of pest control," then shortened it to "biological control" in 1919. The phrase caught on quickly among entomologists.[3]

Smith, who directed biological control work in the state of California from 1913 to 1951, applied the expression only to natural enemies. It did not include pest-resistant crop varieties or agricultural practices later termed "cultural control." Other methods that might be called biological, such as the release of sterile insects, artificial production of insect pheromones, or genetic engineering, did not exist for most of the period considered here. Although Rachel Carson extended the definition to include virtually any method other than pesticides, biological control workers in California preferred Smith's usage.[4] Throughout this study of developments in California, "biological control" will be used in that older, narrower sense, to mean the phenomena and field of study associated with predators, parasites, and pathogens, both before and after 1919.

The term *parasite,* as used in biological control, also has a specific meaning. Most biologists consider a parasite to be an organism that lives in or on another organism and takes sustenance from it but usually does not kill it. Insects that parasitize other insects differ from the typical parasite in significant ways. Insect parasites fall into the same taxonomic class as their hosts and ultimately kill them. These insects, mostly members of the orders Diptera and Hymenoptera, are parasitic only as larvae. They develop from eggs laid in, on, or near their hosts by free-living adults. Unlike typical predators, an insect parasite requires only a single "prey" individual. The developing larva does not hunt as a predator. The adult that finds the host does not consume it. The term *parasitoid* is sometimes used to distinguish insect parasites from either predators or true parasites. Its use is gradually increasing, but is far from universal. Entomologists have long recognized the biological differences, but most continued to use *parasite* throughout the period concerned.[5] This book will follow their usage, which should be unambiguous because true parasites do not enter the story.

The nineteenth-century expansion of American agriculture, and associated growth of government support for agricultural science and education, spawned a new profession called economic entomology. Plagues of insects appeared after farmers planted vast acreage to monoculture. Old-fashioned, manual methods of insect control—such as simply picking off the bugs and squashing them—became impractical in large-scale, commercial farming. Often emerging from farms themselves, eager entomologists combined their fascination for insects with a spirit of service to agriculture. Through farm journals and agricultural societies, and eventually from government positions, entomologists sent farmers the message that science could stop increasingly troublesome insects.

Beginning in the 1850s, a handful of states established professional positions for insect study by hiring official state entomologists. The federal government employed an entomologist even before the creation of the U.S. Department of Agriculture in 1862. Federal activity increased in the 1870s with the appointment of Charles Valentine Riley as chief entomologist. Riley sought no less than a virtual scientific empire. Most states began to support entomology soon after the Hatch Act of 1887 financed a nationwide network of state agricultural experiment stations. Riley helped to launch the American Association of Economic Entomologists in 1889. By the 1890s, Americans claimed world leadership in the field.

Most applied entomologists of the founding generation were self-taught. Some began to teach the subject formally after securing appointments to land-grant universities. The obvious need for insect control helped to sustain economic entomology as a profession that proudly united science and service. Farmers looked to entomologists for answers. The first answers to emerge were chemicals, including arsenic compounds derived from pigments and dyes and bearing the exotic names Paris green and London purple. Concoctions such as kerosene emulsions and soap washes also appeared in the latter part of the nineteenth century. These remained the leading insecticides until other arsenical poisons supplanted them after 1900.[6]

It is difficult to specify why or even when economic entomology came to emphasize insecticides over other methods. Some farmers were dragged into the chemical age before the turn of the century, once the arsenical poisons showed promise against insects such as the cotton boll weevil. Especially in the USDA, more entomologists forsook nonchemical methods by about 1920, after large-scale projects against the gypsy moth and boll weevil failed to achieve rapid control.

Such failures exposed the need for long-term research, while public pressure forced government entomologists to emphasize immediate results. Chemical industry grants and curriculum reform in the 1920s and 1930s solidified the insecticidal orientation of the profession. Biological control also suffered from wartime cutbacks and the appearance in the 1940s of new synthetic organic compounds, including DDT. Various historians have emphasized each of these factors and time periods.[7] Each increase in the attention paid to insecticides was important, but the overall impression is of a gradual movement crowned by the arrival of DDT.

Despite the burgeoning of chemical methods, biological control persisted in California as in no other state, though professional entomologists of any persuasion went there somewhat late. The state had no corps of "bug men" before USDA agents introduced natural enemies from Australia to control *I. purchasi* in 1889. After this triumph, state government officials enthusiastically carried out a biological control program more on the strength of politics than on scientific expertise. They feuded with the USDA until the state brought in entomologists whose credentials satisfied federal authorities. The program secured an academic footing by a transfer to the University of California in 1923. The university maintained, and expanded, an autonomous unit separate from general entomology and thus separate from insecticides. It became the only department of biological control in the nation.

This study will explore the development and activities of California's unique program. Agricultural historians, who have long focused on midwestern grain and southern cotton, tend to neglect the nation's most productive and diverse farming state. Histories of chemical control and the more recent attention to "integrated pest management" allude to California and frame a question: What made California different? Possible causes include the particular crops and attendant insects in California, climate and other environmental circumstances, the organizational structures of California agriculture and entomology, or a special faith that Californians held in a method for which farmers and scientists elsewhere saw little use. As we shall see, all of these factors contributed. Entomology in the USDA provides a contrast. There biological control declined until little was left by the time DDT entered the scene.

Entomology contributed mightily to the twentieth-century trend of standardization in American agriculture. New means of pest control promised to reduce cosmetic damage to farm products. Increased regional specialization and long-distance marketing sparked farmers and middlemen to devise grading scales, largely on the basis of appearance. Standard varieties, packing boxes, and sizes came to prevail. Government-sanctioned grades and commercial brand names emphasized characteristics by which consumers could make comparisons on sight. Producers had to conform to rigid standards to make a profit in a highly competitive market.

Biological control, with its emphasis on "natural" forces, may seem the antithesis of modern, industrial agriculture. It is no coincidence, however, that this method developed most fully in the very state that epitomized industrial farming. Biological control owed its status in California to a specialized, profit-oriented segment of agriculture: the citrus industry. California citrus producers were not stubborn, traditionalist farmers doubting that scientists had any real assistance to give them. They had no tradition to follow.

The farmers and the systems they built were new to California and were business-oriented from the start. They marketed their fruit cooperatively and, in 1908, introduced Sunkist, one of the most successful American brand names for fresh produce. Citrus growers adopted scientific advancements more readily than farmers elsewhere. They tried all manner of new procedures, including pesticides. The most important factor, which above all others made California unique in its support of biological control, was the singular, stunning success against cottony-cushion scale in the industry's moment of need. Timing could not have been better. Nowhere else in the United States did farmers or scientists experience such a dramatic illustration of the power of natural enemies. To a significant extent, the history of biological control in California unfolded as the consequence of this crucial event.

Citrus growers were not alone in their progressive approach. The same businesslike attitude guided other newcomers to California who developed a variety of specialty crops. Eager for whatever help agricultural science could provide, these farmers quickly followed the lead of citrus in marketing and standardization as well. The reluctance of nineteenth-century farmers across the United States to accept scientific advice has probably been exaggerated. Nevertheless, Californians most readily discarded tradition. Their willingness to try new techniques and to organize themselves led them to create official institutions that embraced, fostered, and even forced pest control.

After the first success of biological control saved the citrus industry in the 1880s, growers used their political and economic power to encourage the state, the university, and the scientists themselves to make further efforts. Citrus producers' enthusiasm for biological control led to the specific inclusion of nonchemical methods in the state's new agricultural bureaucracy. This structure protected biological control even from its own failure to get results. Institutional autonomy allowed the method to survive because pressure for short-term gain could not make the group switch to chemicals. The lack of such autonomy paralyzed biological control in other states and in the federal government. Accidents of timing combined with real differences in attitude to prevent biological control from disappearing in California when it did elsewhere.

The citrus industry maintained intimate ties to the biological control program in both state government and the university. In return, the program gave top priority to the needs of citrus. The scope expanded to other crops in a broad and permanent way only in the 1940s. Biological control required keeping a pest population in the field in order to maintain natural enemies. California citrus growers, working in a competitive, long-distance market, imposed cosmetic standards so strict that no insect damage could be tolerated. How and why citrus culture and biological control stayed together will be a recurring theme in this book.

The entomologists considered in this study did not find their practical, agricultural agenda a burden. Though they wished to uncover basic scientific phenomena, they cheerfully placed this desire in the context of a practical mission. Application determined research topics and posed specific questions whose answers often required fundamental research. But application was not an excuse for doing science; rather, basic science was a means toward agricultural ends. Those ends included the blemish-free standards promoted in the citrus industry.

Much of the criticism leveled since the 1960s at agricultural science, and in particular against the emphasis on pesticides, concerns the relationship between scientists and agricultural industries. The most severe critics accused agricultural scientists of abandoning the public interest to serve only those most profit-minded farmers willing to use dangerous chemical technology. Yet entomologists working in biological control shared the same motives and clientele as their counterparts in chemical control. Far from offering a noble, environmentalist alternative to industrial-scale farming, biological control developed within an agricultural science establishment dedicated to all-out production and maximum profit.

From the 1920s onward, the science of biological control became involved in larger theoretical issues in ecology, especially population dynamics. The aim of biological control was to reduce pest populations, permanently if possible. Its practitioners inevitably examined the forces determining animal numbers. Theories of population dynamics grew out of biological control work and came into vehement dispute among ecologists, including both those directly concerned with practical results and those dedicated solely to basic research.[8] In particular, Harry Smith and his colleagues argued for the predominance of density-dependent regulation of populations by biotic forces, including natural enemies. This theory is rooted in the notion of the "balance of nature." I will argue that Smith's position as an applied scientist and the circumstances of his work in biological control deeply influenced not only his theoretical stance but also his very limited participation when the debates heated up. Historians of ecology must consider such contexts when exploring the density-dependence controversy and indeed the fate of the idea of balance itself.

For the California group to be discussed here, and also for USDA entomologists, agricultural application took precedence over pure science. The Californians' theoretical positions defended existing policies regarding the importation of natural enemies. Biological control work produced observations with which the scientists sought to reconcile theory. More than that, however, policy and practice themselves guided the development of theory. Influence rarely moved in the other direction. Biological control practice in California remained largely a matter of trial and error, at least through the 1960s.

Applied biological control led the California entomologists into other areas of more fundamental biological science. Taxonomic study of both pests and their natural enemies was always important in the search for new biological control agents. Mistakes in systematics were blamed for long delays in biological control. Problems in the separation of species and the meaning of designations below the species level drew Smith's group into issues central to the evolutionary synthesis emerging in the 1930s. Alluding to the new methods and theories relating genetics and evolution, and to the Californians' early experience of evolved resistance to insecticides, Smith forecast the eventual failure of chemicals to solve pest problems permanently. Historians of biology have much to learn about the relationship between applied science and the evolutionary synthesis. An absence of strong connections between the two may explain why most economic entomologists failed to heed Smith's warning.

This is a story of science with a mission. The entomologists found it more important to take quick action to save crops and profits than to make long-term studies of how nature worked. Biological control was not easy. In three-quarters of a century it solved only a handful of pest problems. Entomologists in other states and in the USDA concentrated on methods that looked simpler. California did not neglect pesticides—far from it. But California did support a team dedicated solely to the biological approach. Having only that purpose, the team would not turn away when the road looked rough. People seeking natural enemies in the early days boasted of having the only useful method to control pests. Later workers could drop such claims without abandoning biological control.

At least one prominent historian has suggested that American agricultural scientists turned more to basic research after 1900 and felt less need to justify it with promises of practical results or to have much direct contact with farmers.[9] American economic entomology, and California biological control in particular, did not fit this generalization, at least through the first half of the twentieth century. Entomologists identified with the interests of farmers and sought directly to aid that group by reducing the cost of pest control or the loss due to insect damage. Rather than challenge the merits of produce standards based on appearance, scientists faithfully endeavored to meet those standards.

The scientists' mission was agriculture, not environmental safety or public health. Despite his warning about insecticides, Harry Smith should not be considered an environmentalist. Concerns that emerged in the 1960s and led to calls for more attention to biological control were not factors in the earlier development of the California program. Even after Rachel Carson brought biological control to the public eye, entomologists who knew the method best were reluctant to adopt a new constituency if that might mean alienating the old. Growers and scientists alike shared the same goal: to make a spotless orange.

To Make a Spotless Orange

The Scourge

California orange groves were the setting for what became, quite literally, the textbook case of biological control. Any book on the subject will tell the tale of the control of cottony-cushion scale (*Icerya purchasi*) in 1889. Prior to this, entomologists made frequent suggestions and a few actual attempts to introduce biological agents from one place to another to control pests. None of these efforts, however, relieved farmers of damage to their crops or caused any startling disappearance of the target species. The picture for what is now called biological control changed forever after a fluffy, white, Australian insect threatened the survival of California's young citrus industry.

Agents of the U.S. Department of Agriculture (USDA) found insects that fed on cottony-cushion scale in Australia, brought them to California, and spread them through the citrus belt. Eagerly anticipating the results despite the fact that this method had never accomplished much before, growers watched the pest virtually disappear in a matter of months. The success of the vedalia beetle (*Rodolia cardinalis*) and a lesser-known parasitic fly against *I. purchasi* raised hopes—later to be described as fantasies—that all insect pests could be controlled in this way, quickly and completely. The incident sparked a bitter rivalry between California agricultural officials and the USDA. The feud cast a long shadow on the method and its practitioners.

Real life, of course, would not be as simple as early advocates hoped. Later chapters of this book will show that biological control faced a complicated and often frustrating future. Scientists, farmers, and politicians would struggle with the practical use of natural enemies, theoretical explanations of successes and failures, and the very concept of pest control.

The historical import of this single event cannot be overstated. Cottony-cushion scale was more than a good beginning to a story. It became a central part of the folklore of both biological control and the citrus industry in California.

Discussion at every level—scientific, political, or popular— returned sooner or later to the cottony-cushion scale and its famous predator. The first success of biological control was high on drama, adventure, and intrigue. The personalities and ambitions of major participants, combined with the peculiarities of the insects involved, made certain of that. Both scale and beetle were quite conspicuous and indeed beautiful insects. They and their equally colorful history in California made for ready reference among scientists and laymen, whether the subject at hand was research funding, pesticide contamina- tion, or theoretical population dynamics.

T
he idea of using insects to fight insects was not new when cottony-cushion scale threatened California in the 1880s. For centuries the Chinese had encouraged the movement of predacious ants in orchards. Swedish biologist Carolus Linnaeus gave the idea a place in Western science. In what he called the "economy of nature," entomophagous insects filled a crucial role, balancing the numbers of other species. Linnaeus advocated the artificial use of natural enemies. Authors of nineteenth-century European entomology texts wrote that predacious and parasitic insects should be protected or artificially propagated.[1]

W. Conner Sorensen has argued that American farmers accepted economic entomology because of its readily apparent scientific basis in the "balance of nature." Agriculture caused pest outbreaks by upsetting this balance. Agricultural entomologists, even those particularly interested in evolutionary theory, continued to frame pest control solely in this context of balance, which farmers could understand and support.[2]

Biological control passed through its formative period without complex theory because simple theory sufficed. The Linnaean concept, in which each species had others that preyed upon it and prevented it from becoming too numerous, needed little elaboration in an ordered world. Evolutionary theory disrupted the notion that this order was unchanging, but did not overthrow the economy of nature. Indeed, evolution made sense of the profusion of similar species and the often narrow specificity with which some species preyed on others. No biologists observed these phenomena more closely than did entomologists. Insects ate great quantities of each other. Insect species numbered in the millions. A list of natural enemies adorned the typical description of an agricultural pest, although an entomologist could offer little advice on encouraging them. *How* predators or parasites could hold a pest down to a low population density, while depending on a continued supply of food for their own survival, did not become much of an issue until people tried—and usually failed—to manipulate the relationship.

Early attempts to use natural enemies achieved no striking results. In the nineteenth century, however, new agricultural conditions stirred calls for action. Extensive monocultures in American farming created ideal conditions for pest outbreaks. Entomologists recognized that a disproportionate share of the pest species, eventually calculated to be about half, had immigrated from other continents. Most were not known as pests in their native lands. New York's Asa Fitch was the first to blame outbreaks on the absence of natural enemies that would otherwise keep pest populations under control in the Linnaean sense.

"We have received the evil without the remedy," Fitch wrote in 1859. He tried to obtain parasites from Europe by correspondence, but received nothing. Benjamin D. Walsh, a British immigrant to Illinois who turned to entomology at about this time, wrote numerous tirades against patent remedies and chemical cures and called for government

funding of beneficial insect introductions. Walsh scorned Fitch for
naively expecting European scientists to drop what they were doing in
order to locate parasites and supply them to America without compen-
sation.[3]

Walsh thrived on controversy. As one of the early American
advocates of Darwinian evolution, he battled to convert other
entomologists in the years before his death in 1869. It pleased Charles
Darwin to have the argumentative Walsh in the field. Walsh had no
qualms about attacking his own public. He would tell farmers who
misidentified beneficial insects as pests, "You never made a greater
mistake in your life."

He did not draw on Darwin to make a case for natural enemies,
although Darwin spoke to the issue. Darwin observed that most
species were rare despite great powers of reproduction. He pointed out
the essential role of natural enemies in preserving natural balance,
though their effect might go unnoticed until a species was introduced
to a new location without them.[4]

Walsh cited a second reason for the American problem of exotic
pests. He believed that life forms in the New World were more
primitive than those in the Old, an opinion derived from eighteenth-
century French naturalist Georges-Louis Leclerc, Comte de Buffon.
According to Walsh, foreign insects wrought such havoc in America
because European species inevitably defeated their American counterparts. Only European parasites could stop European pests.[5]

Although Walsh wielded a mighty pen, he seems never to have
attempted biological control himself. He talked his way into the
position of Illinois state entomologist by 1867, just two years before
a train hit him and ended his life.[6] Walsh did live long enough to
inspire another English immigrant, who would make the first practical
success with natural enemies.

Charles Valentine Riley, born in 1843, moved to the United
States at age seventeen and worked on an Illinois farm for three years.
A gifted illustrator, he turned his childhood interest in insects into a
career. As an entomological correspondent for the *Prairie Farmer* in
Chicago, he came under the influence of Walsh. After the Illinois
legislature made Walsh state entomologist, Missouri took his recom-

mendation and appointed Riley to a similar post. At this time the only other state entomologist was the aging Fitch in New York. Walsh and Riley continued to work together until Walsh's death.[7]

Although he adopted many of his mentor's views, Riley was no mere extension of Walsh. Riley advocated chemical methods as well as biological. He did not share Walsh's burning passion to import natural enemies, although he agreed that such efforts would be the ideal approach to exotic pests. In his annual report as Missouri entomologist in 1869, Riley copied Walsh's argument for the superiority of European insects and added, "If the American creation is somewhat old-fogyish, that of Australia is the very concentrated essence of old-fogyism itself."[8]

Riley's interests went well beyond routine discussions of Missouri pests. Like Walsh, he pursued the study of evolution. They published together on mimicry, and Riley later wrote on the evolution of the specific relationship between yucca plants and their pollinating moths. He explicitly accepted the inheritance of acquired characteristics as well as natural selection. By the end of his career he gave a more evolutionary explanation for the preponderance of European species among pests in America. Riley did not distinguish between Lamarckian and selectionist mechanisms. He argued simply that European insects had adapted to agricultural conditions through ancient association with cultivated plants. His own experience forced him to give up Walsh's (and Buffon's) theory of European superiority. Riley won his greatest practical triumphs over an American insect that became a pest in Europe, and an Australian insect that became a pest in America.[9]

Riley aspired to be not just Missouri's entomologist, but the nation's. In his nine yearly Missouri reports he regularly decried the national state of entomology, as Walsh had often done. Riley urged more federal government support for the science as a crucial service to American farmers. In 1876 the Interior Department named him to head its U.S. Entomological Commission to investigate the Rocky Mountain locust, which had plagued the Plains and Midwest for several years. The locust outbreak receded on its own, but Riley extended the work of the commission to other pests. He was a logical

choice for entomologist of the USDA when that position became
vacant in 1878. Unable to get along with his superiors, Riley quit after
only a few months, but returned when a new president and cabinet
took office in 1881.[10]

John Henry Comstock of Cornell University filled the USDA
post in the interim. One of the few academic entomologists in America
at the time, Comstock was nevertheless largely self-taught in the
discipline. He gave the first entomology course at Cornell while still
an undergraduate, then built a department that led the nation in
producing college-trained entomologists well into the twentieth
century. Although he lamented being replaced in the federal position
by the egotistical Riley, Comstock was probably better suited to
teaching than to political gamesmanship and lobbying. Riley respected
Comstock's abilities and hired a Cornell student, Leland Ossian
Howard, as an assistant after taking the USDA job the first time.
Howard survived both changes at the top and succeeded to the position
himself after Riley quit again in 1894.[11]

Riley experimented with the chemical pesticides of his day and
fought other entomologists over credit for certain applications, but he
looked for opportunities to try the biological method. As early as
1870, while Missouri state entomologist, he distributed parasites of a
common fruit pest from one area to others where he believed they
might not already exist. The parasites were probably already there.
Two years later he suggested moving hibernating larvae of another
pest to open fields, from which any parasites could escape while the
pests starved. Although he produced no evidence, he later claimed that
these practices had materially aided farmers.[12]

In 1873 Riley did what others had only talked of doing:
deliberately introducing a natural enemy from one continent to
another. He sent a parasite of the grape phylloxera (*Phylloxera
vitifoliae*) to France, where that American aphid was ravaging
vineyards. The parasite became established in Europe but did not
reduce vine damage. Although what would later be called classical
biological control failed to solve the problem, another biological
solution did the job. Riley helped to introduce American rootstock
resistant to the phylloxera. The action saved the French wine industry.

Riley first tried to introduce a European parasite to the United States in 1875. His early attempts to establish *Apanteles glomeratus* against the imported cabbageworm (*Pieris rapae*) failed, but as federal entomologist he finally succeeded in 1883. Like his previous efforts to manipulate beneficial insects, this introduction achieved unspectacular results, if indeed it was of any practical value at all.[13]

By 1887, when an uninvited guest insect called the cottony-cushion scale had the California citrus industry in an uproar, Riley had more experience in biological control than anyone. "The history of similar cases of destructive insects introduced from other countries," he claimed, proved that introduction of natural enemies would control the scale.[14] His own experience promised no such thing, yet he went to great lengths to try the biological approach in this case. Drawn to cottony-cushion scale by political and personal circumstances and by the appeal of a scientific puzzle, Riley grew ever more fascinated with the insect.

Cottony-cushion scale reached North America accidentally and unnoticed in 1868 or 1869, probably in a nursery shipment from Australia to the San Francisco Bay area. The species was rare and inconsequential in Australia. Entomologists there did not know of it until 1878, when it was described—from New Zealand—as *Icerya purchasi*. Gradually working its way south in California, the scale attracted little attention until it descended upon the sparkling orange groves of Los Angeles and the San Gabriel Valley. The insect was

> an insignificant creature in itself, resembling a small bit of fluted white wax a little more than a fourth of an inch in length. But when the scales had once taken possession of a tree they swarmed over it until the bark was hidden; they sucked its sap through their minute beaks until the plant became so feeble that the leaves and young fruit dropped off, a black smut fungus crept over the young twigs, and the weakened tree gradually died. In this way, orchard after orchard of oranges, worth a thousand dollars or more an acre, was utterly destroyed, the best fruit-growing sections of the state were invaded, and ruin stared the fruit growers in the face.

Such was the threat to the booming new citrus industry. As the pest spread, it resisted all efforts at control. The immigrant attacked many other plants and thrived in parts of both northern and southern California.[15] The conspicuousness of the insect's snow-white egg masses would be crucial to the history of biological control.

Riley learned of cottony-cushion scale by 1878. No specialist in scale insects (Coccoidea), he misidentified the new California pest at first but predicted its dire threat to the orchards. Comstock made extensive studies of scale insects in Florida and California while with the USDA. Assistant L. O. Howard later recalled that Riley, upon regaining the USDA job, took special interest in the group because he wanted to best Comstock. Cottony-cushion scale provided an especially good opportunity. Comstock saw *I. purchasi* in Santa Barbara in 1880 but apparently did not know that the species had reached the main citrus belt to the southeast. Nurserymen may have steered him away from the pest for fear of bad publicity, but within a few years the infestation could no longer be hidden.[16]

Although no one else actually introduced natural enemies to the United States during this period, Riley was not the only advocate of biological control. The suggestion to obtain natural enemies of scale insects—presumably certain armored scales, not the then little known cottony-cushion—appeared in print in California as early as 1879. After the vedalia beetle controlled cottony-cushion scale ten years later, several people claimed to have originated the idea with that pest in mind. They may have thought of it independently, but no claim that they gave Riley the idea can be substantiated. Since his days with Walsh, Riley had considered biological control an obvious strategy against any introduced pest. By 1880 both Riley and Howard considered scale insects in general to be good candidates.[17]

Agitation from California to try natural enemies against *I. purchasi* did prompt Riley to make a special project of the pest. He speculated in citrus belt real estate, a fact that may in part reflect his interest and in part explain it. He craved glory for his accomplishments. If anybody was going to conquer the scourge and save the industry, he wanted to be the conqueror. Fighting over credit began even before the scale was beaten.[18] All parties involved held positions that were to some degree political. Political favor, operating funds, even retention of the jobs themselves depended on pleasing some constituency, be it growers, county governments, state legislators, or Congress. Practical results showed that money was well spent, provided one got the credit for the results. Riley was still trying to generate federal support for entomology. An enormous ego only added to his desire for credit. His rivals in California matched his competitiveness—and some would gladly dispense with the truth when necessary.

The cottony-cushion scale problem required more study than Riley's previous biological control efforts. Solving it, he knew, would bolster his credentials as a scientist and public servant. Existing literature enumerated parasites of the pests in earlier cases, but no parasites had yet been discovered to attack cottony-cushion scale. Riley did not even know where to look, because he did not know whence the pest had come. According to the most basic premise of biological control—whether Linnaean, Lamarckian, or Darwinian—any useful natural enemies would probably make a species rare in its native country. To find them, Riley would have to find a place where the scale existed but only in small numbers.

The prospect of detective work through the entomological literature and correspondence with entomologists in other countries only made the cottony-cushion scale more intriguing. Riley also welcomed opportunities to study overseas. His frequent European junkets, however, led Congress to cut off funds for foreign travel by USDA employees, just when he wanted to go searching for the scale's home. As Howard wrote many years later, "Obstacles as a rule only inspired Riley to further efforts." The travel ban fueled his desire to stop the scale and score a personal triumph.[19]

Riley also pursued chemical control. He would take the victory any way he could get it. He placed assistants Daniel W. Coquillett and Albert Koebele in California in 1885 to investigate other pests, then assigned both to cottony-cushion scale later that year. Already fearing entanglement with California's agricultural bureaucracy, Riley urged his agents to keep their distance from local officials. Koebele discredited Coquillett's work after the two feuded over who was in charge. Riley dismissed Coquillett when funds were cut in the summer of 1886. Coquillett returned to the USDA payroll the next year, but in the interim he helped to devise a technique for fumigating orange trees with cyanide. The treatment became standard for several scale insects, although it would soon be unnecessary for cottony-cushion. To Riley's chagrin, the USDA received none of the credit.[20]

Meanwhile, Riley continued his detective work. He concluded by early 1887 that Australia was the only place where *I. purchasi* was known but rare. He visited California for the first time that April and saw the infestation himself. Citing the ban on foreign travel as an obstacle to a search for natural enemies, he discussed other remedies. The Californians later used that fact against him to claim priority for the biological solution. Riley appeared before a convention of fruit growers and the State Board of Horticulture. In an often quoted speech, he urged that Los Angeles County put up the money to sponsor an expedition to Australia, since Congress refused. A few months later, the county appeared ready to do just that and send Coquillett. By then, however, in doubt of the scale's Australian origin and already afraid California would claim credit for any success, Riley rejected the proposal. Confusion of *I. purchasi* with a related species from Mauritius sidetracked Riley for a while. Upon examining specimens in Paris in October 1887, he finally settled questions of synonymy and origin in favor of Australia. He apparently traveled on his own money.[21]

Parasitic insects would be difficult to ship alive across an ocean, as Riley's initial efforts with cabbageworm parasites had shown. Furthermore, he was now dealing with an insect little known in its own country and with no known parasites. But Riley found a correspondent willing to look. Frazer Crawford, of Adelaide, reported a fly parasitizing *I. purchasi* and sent dead specimens to Washington late in 1887. Crawford also corresponded with a state official in northern California and promised to send live parasites. At Riley's request, Crawford shipped them to Coquillett in the south as well. Whether any descendants from these shipments survived and established the fly in California ahead of the vedalia beetle is uncertain. To move live parasites over long distances in the days before air travel required the sending of parasitized hosts. Crawford used not *I. purchasi* but a related scale that is a less suitable host for the cottony-cushion scale parasite now known as *Cryptochetum iceryae*.[22]

Coquillett and Koebele found live parasites at both sites, but could not see persistence in the field. Both agents suspected that two species of parasite were involved. Riley denied the suspicion, but it was confirmed years later. It is likely that most of the parasites emerging from the shipments were *Cryptochetum monophlebi,* which cannot live on *I. purchasi* and could not have survived in California. After Koebele finally went to Australia and sent large numbers of *Cryptochetum* to Coquillett, the latter reported that they would not attack cottony-cushion scale. Coquillett found *C. iceryae* alive nearly two years later, and it eventually flourished also in the north. If credit still matters, assigning it remains problematic. *Cryptochetum* may have been established from Crawford's shipments in 1888 and have failed altogether from Koebele's in 1889, or the reverse may be true.[23]

Riley always insisted that the parasite played an important role in controlling the pest, although this went largely unrecognized for decades. The consensus is that *Cryptochetum* and the vedalia divided up the territory, the parasite predominating on the coast and the predator inland, but that either could have controlled cottony-cushion scale alone. Because the parasite reached California anyway, uncertainty over the exact arrival date is a mere footnote. However, it illustrates a point later workers in biological control would emphasize repeatedly. They complained that taxonomic study received too little support despite great practical benefit. If parasitic species and their host relations were properly catalogued, the lament ran, they could be obtained quickly when needed for use in biological control. Delays as long as eighty years have been blamed on the lack of such knowledge.[24]

The glory, however, went to the vedalia beetle, *Rodolia cardinalis,* because the red and black ladybeetle was as conspicuous as the cottony-cushion scale on which it fed, and because the beetles were easily transferred from one orchard to another. Growers did not have to wait for natural spread. Riley finally sent Koebele to Australia in search of any enemies of *Icerya,* but especially for more

Cryptochetum. Riley bypassed the USDA travel restriction by getting Koebele attached to the State Department delegation to the Melbourne Exposition in 1888. Another entomologist would go a few months later and actually report on the agricultural features of the exposition, while Koebele searched Australian orchards for the scale and its enemies.

Koebele discovered the vedalia in only a few weeks. He shipped several dozen beetles at a time to Coquillett, who released them on an infested orange tree under a tent. On the way home, Koebele stopped in New Zealand to investigate reports that cottony-cushion scale had vanished from areas heavily infested a few years before. He got a preview of what would soon occur in California. *Rodolia cardinalis* must have been introduced to New Zealand by accident within the previous five years. Koebele found the predator sweeping the scale aside. It would become a principle of biological control that the best place to collect natural enemies was where they had just encountered a previously isolated pest population and had not yet diminished their own food supply. Yet the fact that Koebele did it this way would be largely forgotten. Koebele gathered six thousand vedalia eggs, larvae, and adults from New Zealand in three days and brought them to California himself. These formed the greater portion of Coquillett's founding stock.

The beetles multiplied rapidly. Coquillett and growers spread them throughout the infested region, which was virtually clear of the scale within a year. Control has been permanent, except in areas where insecticides have killed the natural enemies and allowed the scale to surge back over the trees. Koebele became the hero of the citrus industry.[25]

Textbook accounts only begin to describe the battle that ensued in correspondence and in the press over credit for importing the vedalia. All of those involved agreed that Koebele, the explorer, deserved honors. Riley also claimed credit for himself as the general running down clues and directing the work. He even named his daughter Cathryn Vedalia. He warned Koebele and Coquillett against allowing California officials to do anything for which they might take credit. Despite his ego, entomologist historians have justly given Riley the recognition he sought. As he feared, various Californians rejected

Riley's assertions for decades afterward and accused him of opposing the entire project.[26]

Among those claiming to have persuaded him to try a search for natural enemies were future state plant quarantine officer Alexander Craw, State Board of Horticulture President Ellwood Cooper, and board secretary Byron Martin Lelong. Cooper and others maintained that they had suggested the Melbourne Exposition dodge to get an entomologist to Australia on federal funds. The idea, however, clearly originated in the USDA. Riley admired Craw and Cooper and did not wish to embarrass them, but he thoroughly despised Lelong, who seems to have held some mysterious control over Cooper. Lelong's publications reveal his dishonesty. After he shot himself to death on the state capitol grounds in 1901, he was found to have embezzled Board of Horticulture funds.[27]

Riley saw the control of cottony-cushion scale as the product of careful research in the literature and in the field by insect experts. His California rivals believed that they had come upon a simple way to defeat pests and glorify themselves. During the next quarter century, the Commission of Horticulture in its various incarnations promoted biological control with grandiose claims and little practical success. Honoring explorer Koebele with both praise and material rewards, Ellwood Cooper called natural enemies God's provision for insect control. "Any other remedy," Cooper wrote, "is expensive, has to be repeated, is unsuccessful and idiotic."[28]

Other remedies would sometimes prove to be all of those. Nevertheless, the method that neatly dispatched one scourge would not be so easily applied to every pest in California. An objective observer—had there been any—might have said some forty years later that the vedalia beetle was a happy accident, or perhaps an unfortunate one because it led government and farmers to waste time and money in vain efforts to repeat the accident. For it would be forty years before Californians defeated another pest as decisively as they had beaten *Icerya purchasi*.

Enthusiasm for biological control survived those lean decades in large part because of the unique hold romantic tales of the vedalia had on growers and politicians. No one who had seen the snowy infestation of scale strangling the orange trees would forget it. The colorful beetle that bred so rapidly and vanquished the pest so easily would likewise remain a vivid memory. Occasional flare-ups, soon found and snuffed out by the natural enemies, reminded those whose fortunes had been saved, and introduced the former pest to growers who arrived afterward. The fact that growers participated directly, carrying insect-riddled branches from one grove to another to spread vedalia beetles, sustained interest in the method. Growers could see it work, which would not always be true of biological control in the years to come.

Scientific and political reputations were built on cottony-cushion scale. Entomologists of a later generation would use the behavior of *I. purchasi* populations to build or verify general theories on population growth and control when few other concrete examples were available. Pesticide-induced outbreaks of cottony-cushion would provide some of the most striking examples of the down side of chemical entomology.

The state's infant agricultural bureaucracy would emerge from this episode with newfound power, and would mature in succeeding decades. Although hyperbole often prevailed over scientific responsibility, this bureaucracy nurtured biological control and established a program that would remain foremost in the world through the mid-twentieth century. To understand biological control in California, one must understand the development of agricultural politics and, especially, the state's powerful citrus industry and its support for research. These movements will be explored in the following chapter.

CHAPTER 2

Citrus, Government, and Science in California

Practical biological control began with cottony-cushion scale in the citrus orchards of southern California. The episode fired the imaginations of farmers, politicians, and scientists in a specialized area of agriculture. Remembering their seemingly miraculous rescue from one pest, orange and lemon growers threw their support behind efforts to control other insects in the same way. No segment of American agriculture matched the enthusiasm and tangible assistance that the California citrus industry gave to biological control in the decades that followed the first demonstration of "bug vs. bug." Although some interest arose in other states, it did not extend to sponsoring a team of scientists whose sole purpose was to fight insect pests with natural enemies.

The entomologists who kept biological control alive in California shared the agricultural outlook of citrus growers. Profit was the name of the game, and the scientists played it without reservation. Insect pests could reduce production but could also cut profits in another way, by lowering the market value of perfectly edible fruit. Rigid cosmetic standards emerged in the citrus industry as it fought for a larger place in the nation's diet. The purpose of pest control was not only to maximize production, but to keep prices high through aesthetic appeal. The purpose of biological control was to accomplish this at minimal cost. Scientists made themselves a component of the industry, and took its goals as their own.

The bond to citrus shaped and sustained California's biological control program. Its industry outlook merits close attention. The development of biological control cannot be fully explained without an understanding of California citrus. The growth of the industry and its ties to both political and academic agricultural science, combined with the remarkable events surrounding the cottony-cushion scale, provided a unique context for this specialized field.

Close association with a specific agricultural industry gave the biological control staff in California a following of enthusiastic collaborators. The U.S. Department of Agriculture, which directed the first successful project, failed to maintain such long-term contact between farmers and biological control researchers. Federal entomologists interested in natural enemies had to consider the whole of American agriculture, but were too few to give any one part of it the kind of attention the Californians gave to citrus.

The defeat of cottony-cushion scale fortuitously brought citrus growers together with biological control in its infancy. The industry maintained a close relationship because scientists and politicians cultivated that support. Citrus was an ecologically appropriate crop for biological control. More important were the particular circumstances of citrus, politics, and agricultural science in California. Those circumstances generated a program that not only survived, but grew, during a period of nearly forty years in which it did not completely control even one more species of insect pest. Emerging from the naive enthusiasm of a few bureaucrats, the program moved toward scientific respectability and was eventually welcomed into the agricultural science arm of the University of California.

From the first stirring of a commercial citrus industry in the state, California growers were more organized and more open to scientific innovation than typical farmers elsewhere. In addition, because citrus growers were producing a new crop at a great distance from the market, their business depended on intense promotion. This inevitably drew their interests into the political arena. Nearly all citrus pests were of foreign origin, as was the crop itself, so the industry sought and received legislative action to stem the tide of insects coming in from overseas. The bureaucracy established for this purpose also dealt with pests that had already reached the state. It was only natural for these officials to attach themselves to biological control, which dealt specifically with foreign insects and lent itself to popular appeal.

Like the politicians, agricultural science administrators of the university aimed to please their constituents. Orange and lemon growers demanded research on their crop. They successfully lobbied for an experiment station devoted to citrus. Government officials had the field of biological control to themselves, however, because Berkeley lacked an entomologist before 1891. Even after that, the university avoided biological control until a bitter rivalry with the bureaucrats began to fade. Upon finally taking over research on natural enemies in 1923, the university bowed to the growers' wishes and placed the work at the Citrus Experiment Station so emphasis on that crop could continue.

T he group of fruits that were to generate such wealth and power originated in Asia. Spanish missionaries introduced oranges to California in the eighteenth century. A former Kentucky trapper planted the region's first commercial citrus orchard in Los Angeles in 1841, seven years before the United States acquired the territory from Mexico and gold was discovered in the north. New settlers in southern California tried a number of crops in search of a profitable industry. The absence of stubborn farm traditionalism was apparent from the beginning. Citrus came to the fore only after the so-called Washington navel orange arrived in the mid-1870s.[1]

A town destined to be the center of citrus research and the institutional home of biological control gained the early advantage. Twenty-five families from the East and Midwest started the colony of Riverside in 1870. Even those moving from farming regions—including a large contingent from Iowa—were mostly townspeople, not

farmers. The colonists came with education and soon secured the
capital to begin an irrigation canal. They planted a few orange groves
almost immediately.

At that time the principal activity of the U.S. Department of
Agriculture was the dissemination of new plant varieties to farmers
around the country. Sometime between 1873 and 1875, a USDA agent
sent two young navel orange trees to one of Riverside's founding
families. The navel originated as a "bud sport" (a somatic mutation)
in a Brazilian orchard, but came to be called the Washington navel
because California got it from Washington, D.C. The first fruit
appeared in 1878 or 1879. Wood from these two trees, budded onto
other rootstock, produced all of the navel orange trees in California.

The new orange variety thrived under irrigation in the warm,
well-drained, upland area of Riverside. In a few years, grafted trees
spawned a belt of citrus towns from Pasadena to Redlands. Several
towns originated as colonies like Riverside. Riverside navels received
national acclaim at fairs, including at least one the industry's own
promoters convened. Citrus cultivation boomed with the connection of
the transcontinental Southern Pacific Railroad to southern California
in 1876 and, more important, the arrival of the Santa Fe line and
competitive fares in 1885. Part of the larger picture of regional crop
specialization in America, the railroads and their new refrigerated cars
allowed growers to build a nationwide market for California fruit.
New lemon orchards in Ventura Country enlarged the citrus empire in
the 1890s. After the turn of the century, increased production of
summer-ripening Valencia oranges in cooler coastal areas such as
Orange County complemented the winter navels of the interior to give
California a year-round crop. Grapefruit trees were planted later in the
desert valleys farther inland.[2]

Books promoting the climate and opportunities in California
lured new settlers with promises of a middle-class living from a few
acres of trees that would bear golden fruit for years after initial
planting. People with a background in business, rather than agriculture, sought a suburban life in the citrus towns—although the towns
have since degenerated into urban sprawl as houses and shopping
centers displace the groves. The orchards were actually in the towns,

not in outlying rural areas. New residents could build the schools, churches, and other cultural amenities to which they had become accustomed elsewhere, and still live on their "farms."

Indeed it was possible to make a living from as little as ten acres—if one had the right ten acres. But retired or active doctors, lawyers, ministers, teachers, and other professionals owned many of the small orchards as a sideline. Large operations dominated the economics of citrus, and the smaller growers went along. Increased values of suburban farmland ensured that new growers would continue to come from the well-to-do classes. Too many would-be farmers came, and went bust learning that only the parcels with ideal soil, drainage, temperature range, and other conditions could provide a successful livelihood in the highly competitive new industry.[3]

The businesslike, suburban growers present a vivid contrast to the established picture of the recalcitrant, nineteenth-century dirt farmer, who clung to tradition and resisted scientific innovations. Historians have argued that, as agricultural science became professionalized and institutionalized in state experiment stations, scientists strove to professionalize farmers as well. But the "reluctant farmer" did not grow oranges. California citrus producers were professional businessmen from the start, with no tradition to which to cling. They dismissed European precedents for citriculture and adapted any available information to California conditions.[4]

California orange growers' experience strikingly parallels the emergence of the prairie rice industry in Louisiana in the 1880s. Progressive farmers from the Midwest, especially Iowa, found small-scale rice farming among the Cajuns and transformed the industry with new methods, new machines, and, in the late 1890s, a new variety introduced from Japan and especially suited to Louisiana prairie conditions. As in California citrus, rice growers formed successful cooperative associations and increased the national demand for their product.[5] Even within California, while the suburban form of agriculture perhaps was peculiar to citrus, the business orientation of newcomers was not. Eager innovators of the same period sought to build fortunes in products ranging from sugar beets to wine to silk, and led the nation in mechanizing wheat farming.[6]

Lacking tradition, California growers eagerly sought anything that agricultural science could do to increase their profits. Like farmers interested in science elsewhere, Californians wanted practical information immediately. Agricultural scientists at first found it difficult to please both farmers and university administrators. The former desired quick results, while the latter stressed the academic side and were wary of agriculture. The University of California fired its first professor of agriculture in 1874 for being too practical. His successor, geologist and soil scientist Eugene W. Hilgard, won the growers over while satisfying the board of regents with his scientific standing.

Historian Alan I Marcus has portrayed agricultural scientists of the period as seeking to assert authority over farmers, including the group he called "scientific farmers" who wanted scientific advancement but expected to produce the advances themselves. He included Hilgard among those scientists who considered their chief educational role to be training new scientists rather than farmers. Although Hilgard was an accomplished basic scientist, he does not fit Marcus's model. Hilgard emphasized his experiment station's service role and addressed the applied problems directly relevant to farmers, especially state soil surveys. He even clashed with USDA policy makers insistent on basic research. In his eagerness to turn funds from the Hatch Experiment Station Act of 1887 to practical ends, Hilgard opened four branch stations that the USDA later forced the university to close.[7]

Far from opposing Marcus's "scientific farmers," Hilgard urged citrus growers to continue experimenting and exchanging information and to be their own experiment station. In the citrus belt the university cooperated with Pomona College entomologist Albert J. Cook, who organized dozens of horticultural clubs in the 1890s and 1900s to encourage such activity. California farmers as a whole welcomed agricultural science as long as they had some indication of practical results. None were more willing to listen than the citrus growers, transplanted businessmen already accustomed to receiving advice in print.

The university came to growers through farmers' institutes, local lecture series that served as an early form of agricultural extension.

Hilgard's staff also freely used the state's leading farm journal, the *Pacific Rural Press*. *Press* editor Edward J. Wickson, a leading promoter of California and of citrus, taught in the College of Agriculture and later succeeded Hilgard as dean. Hilgard established a formal agricultural extension program in 1903. Perhaps the numbers of actual experimenters were not much greater in California than elsewhere, but California citrus growers eagerly followed those who would lead.

Those who would lead took advice from each other or from any institution willing to provide it. They listened to journal writers, university scientists, state agricultural officials, or the USDA, although these groups often offered conflicting advice. Most advances before the turn of the century originated among the growers themselves, but biological control of cottony-cushion scale by the USDA helped increase their acceptance of institutional science.[8]

Especially in regard to insect problems, the institution that seemed to offer the most to progressive farmers was the network of state and county horticultural commissioners. This system emerged in the 1880s through political appointment. California agriculture—and citrus most of all—depended on enthusiastic promotion, a task for which successful politicians were well suited. Intended to address insect problems, and particularly insects introduced from abroad, the agricultural bureaucracy was the most obvious setting for early interest in biological control. An entirely political institution could only be expected to exaggerate the potential benefits, and it did.

Prior to the establishment of the California Department of Agriculture in 1919, a series of appointed bodies dealt with fruit crops and especially their insect pests. The principal duty of these agencies was to prevent the entry of new pests from other countries. A long coastline and busy ports made California especially vulnerable to the introduction of insects and plant diseases from abroad. Ornamental and agricultural plants imported by nurseries constantly threatened the young state with new pests. Almost no native American insects attacked citrus, a foreign crop still expanding largely through the importation of nursery stock. With so much exotic plant material coming in, it was only natural for California to lead other states in enacting plant quarantine legislation. The state appointed a Board of Viticultural Commissioners in 1880 in response to demands from wine growers, and then a Board of Horticultural Commissioners subordinate to it. An independent State Board of Horticulture replaced the subordinate board in 1883 and took over quarantine inspection.

Ellwood Cooper, a wealthy Santa Barbara olive and walnut grower appointed to the first horticultural board in 1881, presided over the second for most of its history and became sole Commissioner of Horticulture when the service was reorganized in 1903. He remained in office until 1907. The Pennsylvania native had already amassed a considerable fortune in the shipping business before moving to California in 1870. Cooper was no tradition-bound dirt farmer, but an educated businessman like so many of the citrus growers whose cause he took up. He was considered for U.S. Secretary of Agriculture at least once in later years.[9]

A law authorizing port inspection and destruction of infested plants was enacted in 1881, but it proved both toothless and unconstitutional. California passed the world's first effective plant quarantine legislation in the midst of the cottony-cushion scale battle in 1889. Alexander Craw, manager of a Los Angeles orchard and principal southern California champion of plant quarantine, was named state quarantine officer and inspector in 1890. Craw held the job at the port of San Francisco for the next fourteen years. The quarantine officer sought quick eradication of new pests that appeared in the state, before they could spread so far as to make eradication impossible. The new law passed its first major test when Craw seized 325,000 infested orange trees imported as nursery stock in 1891. He condemned the trees as a public nuisance, and the state supreme court upheld his action.

Legislators strengthened quarantine law in 1912 after the Mediterranean fruit fly, considered a dire threat to numerous California crops, was discovered in Hawaii. The California Commissioner of

Horticulture stationed an inspector in Honolulu to stop infested fruit from ever sailing to California. The "medfly" remains one of California agriculturists' greatest fears.[10]

It was also in 1912 that Congress finally authorized the first federal plant quarantines. Other countries had restricted American produce for twenty years for fear of acquiring pests from the United States. The USDA gradually took over international quarantine and left only interstate restriction up to the states. The California Department of Agriculture (now Food and Agriculture) absorbed the Horticultural Commission's quarantine responsibilities. The state still inspects more vigorously than any other, as motorists crossing its borders can attest.[11]

The horticultural commission could not keep all foreign pests out of the state. Some were already established, and others eventually slipped past the quarantine net. The commission tried to stamp out new infestations before they became too large to eradicate. Ellwood Cooper made biological control his preferred means of fighting an insect once it had spread beyond hope of eradication. Quarantine officer Craw did much of the hands-on work, receiving predacious and parasitic insects from abroad and distributing them in California. But officials at the local level made day-to-day contact with growers and with insect problems lacking a ready biological solution.

California's distinctive system of county horticultural commissioners (known since 1919 as agricultural commissioners) evolved in conjunction with the state commission. Under state law, counties appointed boards of commissioners with police powers to restrict intrastate movement of infested plant materials and to force growers to treat infested orchards with chemicals. If growers refused, the commissioners could order treatment and make the expense a lien on the property. This sometimes placed county boards at odds with the state board's policy of introducing and protecting natural enemies of pests. Growers wishing to rely entirely on natural enemies pleaded with state officials to prevent the counties from forcing treatment. The legislature eventually extended the commissioners' powers to destroy abandoned or neglected orchards as potential havens for pests even if the orchards were not then infested.[12]

Appointments depended more on political allegiance than expertise in the early years of the commissioner system. Counties shifted to a single horticultural commissioner in the first decade of the new century. The system was standardized after 1909. Candidates had to pass examinations of their technical competence, under the auspices of the state commissioner of horticulture. Duties later included enforcement of produce and packing standardization laws. California's county commissioners also served some extension functions. They continued to do so even after 1914, when the Smith-Lever Act formally established the federal Extension Service as a cooperative association of the USDA and agricultural colleges.

One way in which the state and county commissioners spread information was through the semiannual fruit growers' conventions they conducted, alternating between northern and southern California. If California growers already welcomed scientific advances more than the average American farmer, those attending the conventions were a still more progressive subset. They discussed pest control at every meeting. Indeed, as one historian has argued, the pest control campaigns that built the bureaucracy gave growers the organizational impetus toward later lobbying for university research and toward cooperative marketing.[13]

During the 1910s, California joined the national trend of states separating research and regulatory functions into universities and government departments, respectively. The California Department of Agriculture, established in 1919, consolidated the state's various agricultural agencies. Pesticide regulation, previously a university responsibility, shifted to the department. The transfer of biological control to the university in 1923 helped complete the restructuring of authority.[14]

Because state government, not the university, directed biological control efforts prior to 1923, contact with growers went through county commissioners rather than extension agents. Ties between the biological control program and the commissioners remained close after the university took over research. Not only were commissioners attuned to the method, but any deliberate introduction of an insect still required the county commissioner's permission.

As state horticultural commissioner up to 1907, Ellwood Cooper came increasingly into conflict with the state's other agricultural science institution, the University of California. A university chemist, among others, developed a process for fumigating orange trees with cyanide gas in the late 1880s. While chemical control soon became unnecessary for cottony-cushion scale, cyanide remained in use against other pests until the 1940s. After the university appointed entomologist Charles W. Woodworth to its faculty in 1891, he championed fumigation and directed research on other chemical control methods. Agricultural college dean and experiment station director Hilgard and his assistant, *Pacific Rural Press* editor Wickson, maintained a cautious attitude toward biological control. Cooper, who touted natural enemies as the only useful means of pest control, regularly denounced Woodworth and the university.[15]

Citrus growers wanted help from both government and university. As they organized their industry, growers became a politically powerful group themselves. They were frustrated by the university's lack of further innovations in citrus production, and particularly the inattention to citrus at Hilgard's Pomona substation. Many felt that he should have located the station in a prominent citrus town like Riverside. They lobbied for an experiment station devoted to their needs. Also flexing new organizational muscle were south coast walnut growers, who sought a university plant pathology laboratory.

Answering both groups in 1905, the legislature authorized a plant pathology lab at Whittier and a soil station emphasizing citrus at Riverside. Work began the following summer. Bowing at last to USDA pressure, the university closed the old Pomona facility—which never had a permanent scientific staff—along with the other three field stations.

Unsatisfied with the productivity of the fledgling citrus station, growers lobbied for more. The new dean of the College of Agriculture, Thomas F. Hunt, began to reorganize the university's entire agricultural program upon his arrival in 1912. He gave citrus immediate attention and named Cornell University plant breeder Herbert John Webber director of an upgraded Citrus Experiment Station with a full scientific staff and facilities. The San Fernando

Valley, not yet a major citrus area but about to receive water from the
controversial Los Angeles Aqueduct, nearly pulled the station away
from Riverside in a renewed political battle. Webber favored Riverside. He won out in 1914 only because of the united support of the
older citrus belt. Residents knew the economic importance of the
single crop in their midst. They toed the line set by the relatively few
powerful growers. Eventually the work of the Whittier lab, including
entomology, moved to the new Citrus Experiment Station on the site
that forty years later would become a university campus. Webber's
station conducted both fundamental and applied research pertaining to
citrus, although other southern crops received attention as the years
went by.[16]

The entomologist assigned to Riverside shared the university's
skeptical attitude toward biological control and concentrated on
pesticides. He had to pay some regard to natural enemies if the
university was to hold its own in the contest for southern growers'
allegiance. Lacking the facilities or the inclination to import beneficial
insects, university entomologists evaluated the effectiveness of natural
enemies already in the field, often disputing the government men's
claims.[17] At about the same time that the Citrus Experiment Station
was taking shape, however, a new generation in the horticultural
commissioner's office began to make peace with the university and
take a more reasoned approach toward biological control. The program
was transferred to the university in 1923, and citrus leaders insisted
that biological control be located at Riverside. By then it was welcome
there.

S uccessful lobbying for university research was only one result
of citrus growers' ability to organize themselves. Their greatest
achievement was the cooperative marketing association now
known as Sunkist Growers. Through it they turned the tables on
greedy middlemen, transformed citrus from a luxury item into a
national staple via advertising, and made the premium Sunkist brand

a household word. The industry raised pest control standards so that it could market spotless fruit nationwide. In their eagerness for practical scientific advancement and for the most cost-effective way to meet the cosmetic ideal they set for themselves, growers supported biological control. They gave it their enthusiasm, their political weight, and their money.

The long distance from their market left citrus growers at the mercy of middlemen who forced them to take all of the risks. Growers paid fixed rates to ship their fruit east and received for it only as much as the buyer, protecting his own profit, deigned to pay. The individual, unable to control when and where others shipped, could not avoid gluts that drove prices down in local markets. As promotion of California continued to attract new growers, and orchards already planted began to bear fruit, the crop grew faster than demand. A Riverside grower organized ten others into an association that pooled fruit and marketed it as a unit in 1892. Similar associations immediately followed this example. Groups of small cooperatives formed district exchanges the next year.

Moving swiftly to establish an organization with real clout, delegates met in Los Angeles in August 1893 and formed the Southern California Fruit Exchange, comprising seven district exchanges. The association retained its three hierarchical levels as it grew. It changed its name to the California Fruit Growers Exchange in 1905 to reflect the inclusion of San Joaquin Valley growers. By then 25 district exchanges, 201 local packing associations, and 15,000 growers had joined the movement. Participation fluctuated widely in the early years, but more than half of California's orange crop and nearly all of the lemons eventually came under the Exchange banner. With its own network of marketing agents in major eastern cities, the organization could regulate the flow of citrus fruit throughout the country, maximize and stabilize prices, prevent local gluts, and keep middlemen from playing growers off against each other. California producers of walnuts, almonds, stone fruits, raisins, and dairy products followed the example of citrus and formed associations by 1915, ahead of successful cooperative movements elsewhere in the United States.[18]

Earlier attempts at cooperative marketing of various crops,

especially by the Grange and the Farmers' Alliances, had failed conspicuously. Citrus was a relatively new enterprise. A businesslike attitude enabled growers to respond to economic crisis with carefully considered measures. They lacked the individualistic tradition that thwarted cooperative ventures in older farm regions. Citrus growers could diagnose their trouble more easily. It was clearly a problem of the long distance market, which they came to know intimately. They did not attach cooperative marketing to any broader, political cause. Their numbers were relatively few, which made organization and staying organized more practical than it would have been in, say, the entire midwestern corn belt. These advantages applied, perhaps to lesser degrees, to the other specialty crop associations in California. The latter groups also had the example of immediate benefits in citrus.[19]

As an organization of separate owners voting by acreage, the Exchange dodged the label of trust but had many of the advantages of a trust. One small grower could not afford to operate a packing house; a local association could. The groups also cooperatively handled irrigation, application of fertilizer, orchard heating, insect control, pruning, and harvesting, using their own permanent staff or contracting the work. Orange trees needed more pampering than the early, eager growers had expected. But many, especially those with smaller orchards, maintained the social distinction of a citrus grower while doing little or none of the actual work themselves. Voting by acreage encouraged large operators to join the Exchange, choose its officers, and make the business decisions, while smaller growers who were retired or whose chief occupation lay elsewhere were content to let the associations manage the business.

This system led to sharp social stratification in the citrus belt. Citriculture required enough work throughout the year that growers or associations could hire workers permanently, but the employees—first largely Chinese, then Japanese, then Mexican—were as far removed from the owners socially as were migrant laborers in other crops. As social critic Carey McWilliams described it, the middle class of townspeople allied itself with the "do-nothings who own the groves," and did not protest when growers brutally put down workers' attempts to organize in the 1930s and 1940s.

The California Fruit Growers Exchange led the antiunion Associated Farmers in southern California, and longtime Exchange president Charles C. Teague articulated the views of the dominant class. In his autobiography, he pointed to the common interest of large and small growers. Teague proclaimed that California agriculture had no class conflict. He then condescendingly described the "natural" attributes of respective ethnic groups. Mexicans, he believed, were "naturally adapted to agricultural work" because they came from a similar climate. Citrus growers were an educated, professional, powerful social class in a segregated society—willing to listen to, and eager to reward, tax-supported scientists who could help them increase their profits.[20]

The scientists knew who their customers were and aimed to please them. Scientists worked for the landowners on the vague assumption that what was good for this group was good for everybody. If growers could cut costs while increasing production, then everyone would have more food on the table. Agricultural biologists and chemists could not anticipate the sociological drawbacks of this outlook. Citrus scientists were a part of the citrus culture. They saw the world through the growers' eyes. Years later, when critics charged that pesticides endangered farm workers and that mechanization took jobs away, the scientists could say that they had only done *their* jobs and done them well. Biological control was no exception. Although nonchemical control might be more socially and ecologically responsible than insecticides, that is not why the entomologists pursued it. They did it to save the growers money.

The spokesman and leader of the citrus culture, Charles Teague, supported biological control with enthusiasm. For a time, Teague wielded more political power than anyone else in California agriculture. The Maine native came to Santa Paula in the 1890s, acquired a 40-acre lemon orchard, and joined it to the Limoneira Company. After becoming manager of the company in 1898, he built Limoneira into the state's largest citrus holding and the world's largest lemon acreage.

Ventura County growers joined the California Fruit Growers Exchange in 1901. Limoneira pulled out a few years later, but rejoined the Exchange for good in 1911.

Teague soon rose to the top of the cooperative movement. He presided not only over the California Fruit Growers Exchange from 1920 to 1950 but also, for roughly the same period, over the state's walnut and lima bean cooperatives. He operated farming and water companies in Ventura County and in the Salinas Valley of northern California. He headed the principal farm lobbying group in the state, the Agricultural Legislative Committee, in the 1920s. President Herbert Hoover named him to the Federal Farm Board. Whenever he desired, Teague had the attention of governors, agricultural officials, and University of California administrators. It was he who made sure that the biological control program continued to give top priority to citrus. He also served as a regent of the university from 1930 until his death in 1950.[21]

Activities of the California Fruit Growers Exchange (Sunkist Growers since 1952) have been many, including byproducts research and the licensing of the Sunkist name for candy and soda. The Exchange went into the lumber business in northern California in 1907 to ensure a cheap supply of packing-crate wood after the rebuilding of San Francisco sharply increased demand. As a result, the growers' intense interest in pest control extended even to pine insects. The subsidiary Fruit Growers Supply Company grew to encompass all manner of equipment, from fertilizer and cyanide to orchard heaters.[22]

Occasionally Sunkist took researchers from public service and put them on the growers' payroll. The most notable case was G. Harold Powell, a USDA scientist who spent five winters in Riverside and solved one of the most pressing problems in the industry. Too many oranges molded en route to the East despite refrigerated cars. Powell found that more careful handling, and precooling the fruit before loading it onto the trains, prevented the decay. He became a hero to the growers, and in 1912 the Exchange hired him as general manager.[23]

The Exchange made its greatest impact, however, by increasing the national appetite for citrus. By 1907, planned distribution to regulate the market had gone about as far as it could in selling a luxury crop. To do more required advertising. The cooperative balked at the idea until the Southern Pacific Railroad offered to split the cost of an ad campaign. The railroad had its own best interests at heart, for in selling oranges the ads could also sell orange growing and attract more immigrants to become fruit shippers in California.

The initial orange campaign blitzed the state of Iowa early in 1908, touting "Oranges for Health, California for Wealth." One stratagem involved giving away a carload of oranges in promotions such as school essay and poetry contests. Orange consumption shot up 50 percent in Iowa that year, compared to 17 percent nationwide. The Exchange spread ads to neighboring states in the second year, and throughout the nation in the third. Riding a wave of stunning success, the growers rapidly increased their advertising budget. They reached the consumer via newspapers, magazines, billboards, and streetcars.

Advertisers touted every health benefit discovered for citrus over the years, including vitamin C, and invented a few more. "Eat oranges; they purify the blood," consumers were told. "Eat more oranges and keep away 'flu.'" In later years, beginning with a University of Chicago study in the 1930s, the organization sponsored nutrition research it could use for advertising. Lemon ads suggested myriad household uses for the fruit, which consumers would not eat fresh. The Exchange distributed citrus recipe booklets. It sold juice extractors at cost to soda fountains to encourage consumption, and made orange juice a daily breakfast habit for millions of Americans, although Florida later won most of the juice market.[24]

Success was almost dizzying. Americans ate nearly 80 percent more oranges in 1914 than in 1907. California orange acreage increased 70 percent in the same period. Promotion of oranges attracted new growers, just as railroad executives had hoped. Local packing associations unwittingly encouraged immigration with their brilliantly colored crate labels, which sold California's image and which have since become collectors' items. Advertising made citrus a national staple, but continued growth of demand could barely keep up with soaring production. Orange consumption grew from nineteen to twenty-six pounds per capita between 1925 and 1935 and took a bite out of the apple market, but the citrus industry still was unable to sell all it produced.[25] The Exchange could not limit production. Until the 1930s Teague and other leaders insisted that overproduction did not exist, only underconsumption resulting from inadequate regulation of the market. They blamed this on independent growers, who shared some of the benefits of advertising while avoiding its costs, and who sold quickly and erratically instead of waiting for orderly distribution by the Exchange.

Teague, of course, continually urged all growers to join the cooperative, but the individual's interests often clashed with those of the group. The association kept average prices high by preventing gluts in any local market at a particular time. A nonmember could sell his or her whole crop in such a protected market without paying for the work that protected it or for the advertising that boosted demand. The greater percentage in the association, the greater average return for all growers, yet the individual who stayed out fared best of all. Any attempt by the Exchange to curtail production in addition to regulating distribution would have increased the incentive to pull out.[26]

The purposes of advertising were to increase consumption of citrus, of California citrus, and of California Fruit Growers Exchange citrus. To focus on the last benefit, which only members could receive, the Exchange introduced the brand name Sunkist for its premium fruit in the second year of advertising. The suggestion came from the advertising agency. Success of a premium brand required consistency to build consumer loyalty, which became ever more important as production increased. The Exchange established a grading system and exhorted members to emphasize "high quality fruit—the kind which will make the consumer want more."[27] Just as nonmembers damaged the industry by disrupting market regulation, shiftless growers who sent to market anything less than top-quality fruit eroded consumer loyalty and tarnished the name of all California citrus. So went the company line, at least.

As far as advertising was concerned, quality meant appearance. Only the prettiest fruit received the Sunkist name. While the organization marketed the crops of all of its members, growers whose oranges were biggest, roundest, shiniest, and free of any trace of insect injury received the highest price. It mattered little that these characteristics did not determine the quality or taste inside a thick-skinned orange. A premium brand sold by cosmetic appeal. In the words of a 1925 insecticide ad, "No matter how good it tastes, looks are what sells."[28] In the Sunkist-dominated citrus magazine, the *California Citrograph,* Exchange leaders constantly reminded readers that one grade level could make the difference between profit and loss. Growers must take all possible steps to prevent damage, and absolutely must not ship damaged fruit. Because top grades rewarded the individual grower, the system restored a degree of competition within the cooperative group. Occasionally, even after decades of success under the brand, some growers grumbled that all Exchange citrus should be called Sunkist. The directors still reserved that label for top grade fruit.[29]

In setting grade levels for their fruit, citrus growers led a movement of agricultural producers toward standardization. It was only at the urging of producers that states and the USDA established legal standards, to protect those farmers able to meet them. Verifiable standards allowed meaningful communication and comparison in the increasingly complex, long-distance marketing of farm products early in the twentieth century. Wholesale buyers could make deals without direct inspection of every shipment. Advertising and brand names such as Sunkist heightened the importance of consistency. By determining a simple ranking for each grower's fruit, standard grades also facilitated the pooling of fruit by cooperative associations.[30]

The USDA began to set legal standards for produce in 1913. More crops came under grading systems after the department initiated shipping and receiving point inspection in 1917. While USDA grades were voluntary, certification insured that a shipment would be accepted at the receiving end. States set parallel standards, but often made them compulsory. Defended by the California Fruit Growers Exchange, specific state standards such as the proportion of soluble solid to citric acid in fresh grapefruit survived court challenges in the 1920s and 1930s. Sunkist growers who met the standards forced others who did not to have their fruit legally labeled inferior.[31]

To keep their oranges spotless and beautiful, growers had to control insects. Even a few tiny red scale or discolored spots caused

by other insects ate into profits, if not into the taste or nutritive value
of the fruit itself. In light of such exacting standards, any support for
biological control might seem surprising, since that method depended
on the survival of some of the pests in order to maintain a population
of natural enemies. Growers sprayed and fumigated intensively, but
they remembered that one of their most serious pests, cottony-cushion
scale, had fallen before the conquering vedalia beetle. Conscious of
the bottom line, growers knew that permanent biological control, once
achieved, would be cheaper than chemicals. Even as spray guns and
cyanide tents moved through the groves, leaders such as Charles
Teague eagerly supported the state's biological control program and
made sure it served the citrus industry.

Cosmetic standards were depicted as protection for the consumer
as well as the deserving producer. Legal measures to reduce spoilage
surely served both purposes. Purely cosmetic standards foisted on the
public, however, reduced the number of varieties of a particular crop
and improved neither taste nor nutrition. The selling of premium
brands by their appearance alone increased the pressure on farmers to
control insects. Visible damage, even if harmless, could not be
tolerated. In later years, when a new generation of pesticides promised
virtual elimination of insects from the fields, governments enacted
stricter standards against the marketing of produce with even slight
damage or tiny amounts of insect parts. But this all happened because
the industry wanted it. Scientists trying to improve pest control—
whether biological or chemical—endorsed the industry view of
standardization and strove to meet the challenge.

The entire agricultural science establishment, in California as
well as in other states, worked to increase farmers' profits. Scientists
aided those farmers most willing to take their advice. Within any one
crop group, the wealthiest, most business-minded farmers with the
largest holdings were the ones most eager for scientific help. Among
different crop groups, none were more eager than California citrus
growers. More suburban than rural, more businessmen than farmers,
they discarded tradition and built a powerful industry, changing
American eating habits along the way. Legislators, bureaucrats, and
university administrators mobilized to serve an industry whose success

they saw as California's success. Scientists found in citrus a constituency that acknowledged their authority and ability to help. The wishes of the producer became the wishes of the scientist.

In California as in no other state, biological control captured the attention of an industry. Having once experienced successful biological control, citrus growers were eager for more. Agricultural officials kept the movement alive through enthusiasm, naiveté, and, sometimes, deceit. After giving the growers an experiment station devoted to their crop, the university took in and nurtured biological control because growers wanted it. They continued to want it because scientists studying natural enemies remained attuned to the needs of the industry. The scientists discussed insects in economic terms, kept in contact with leading growers, and in some cases became growers themselves. They shared, and did not try to change, the goals of the industry. A close association between biological control workers and citrus growers helped the program to endure.

CHAPTER 3

The Parasite Craze

The conquest of cottony-cushion scale in 1889 laid the foundation of an enduring alliance between the California agricultural bureaucracy and the state's burgeoning citrus industry. State Board of Horticulture President Ellwood Cooper and his band deserved far less credit than they claimed, but proximity gave them an advantage over the federal government in the battle for growers' allegiance. For decades, persuasion would be more important than actual achievement.

State officials lobbied the U.S. Department of Agriculture to sponsor more foreign expeditions for natural enemies of California pests. Cooper and his followers argued that natural enemies alone would solve all insect problems. When the USDA balked, the state established its own program on the assumption that "the parasite" of any insect would vanquish it. One had only to find the parasite.

With political backing and the support of citrus growers, but without much knowledge of insects, California horticultural officials pursued this program for twenty years. They denounced anyone who dared claim that the "parasite craze," or "ladybird fantasy," would not succeed. The unrestrained enthusiasm of this period failed to control any additional pests, but nevertheless paved the way for a more reasoned approach that prevailed after 1913. Meanwhile, the Hawaiian government and the territory's powerful sugar producers seized on biological control as the answer to their own entomological concerns.

Exaggeration, accusation, and disappointment marked the period following the real success of the vedalia, but that exaggeration kept biological control alive.

Although economic entomologists intended their work as science in service to agriculture, decades of squabbling in and between Washington and Sacramento failed to solve even one pest problem. The USDA directed the initial project in California, but afterward federal and state officials took contrasting attitudes toward biological control. The two factions remained in contact primarily as rivals. USDA entomologists took the more cautious approach, warning against false hopes and stressing the need for extensive study. The federal department had the better scientific credentials and facilities for the study of insects. It was there that scientists developed a theory to explain the control of a pest by its natural enemies. The Californians, however, cultivated grass-roots enthusiasm. They built a program dedicated to biological control after the fortunate success of one project—and falsely claimed credit for that one.

The federal government spread biological control efforts thinly across its administrative structure. The method did not receive top priority in any division. After failing in one all-out, coordinated effort to stop the gypsy moth, the USDA never matched its scientific respectability to a real team approach. USDA leaders sought a monopoly on biological control, fearing that California would introduce harmful insects by mistake. Federal efforts in the early period were hardly safer than the state's, yet U.S. Bureau of Entomology Chief Leland O. Howard cited the example of California in order to prevent other states from entering the field. Neither system achieved any practical results in the early going. Cooler heads eventually caught up with enthusiasm in California and made biological control realistic and effective.

Each side could pursue biological control with little regard to its theoretical basis. A general notion that nature could keep itself in balance sufficed. Biological control was an effort to restore the instruments of that balance where they were missing, rather than to understand how they worked. Even after specific theoretical ideas began to emerge, the practice of biological control remained largely

empirical. The same was true of chemical control. Economic entomology gained acceptance in the public eye through service, not intellectu-
alism. California's agricultural bureaucracy and the USDA each
claimed to be the true provider of service.

A s vedalia beetles feasted upon cottony-cushion scale all over
the citrus belt, no one appeared to be a greater servant of
agriculture than explorer Albert Koebele. Koebele had made
biological control look easy. After one quick trip to Australia, cottony-
cushion scale ceased to be a pest. His combative boss, Charles
Valentine Riley, knew from experience that such work would usually
be more difficult. Riley ordered USDA entomologists to continue
experimenting with chemicals. Cooper and his board's wily secretary,
Byron Lelong, wanted to send Koebele back to Australia after enemies
of California red scale (*Aonidiella aurantii*) and other citrus insects.
Koebele's report that red scale was as much a pest in Australia as in
California could not sway the Californians from their belief that
Australia would meet their every need. In the ensuing years, they at
least came to realize that pests could come from anywhere.

Riley refused to authorize another voyage without sufficient
evidence that it could solve a particular pest problem. If the state
horticultural commissioners insisted on another expedition, he hoped
that they would employ an agent of their own rather than tie up one
of his men. Lelong and Cooper went over his head and convinced
Secretary of Agriculture Jeremiah M. Rusk to force Riley to send
Koebele back to Australia, on a USDA salary but with California
paying expenses. Cooper and his staff would later claim that the
secretary privately ordered Koebele to report to them and ignore
Riley's instructions. Records indicate, however, that Rusk left Riley
in charge despite the entomologist's opposition to the trip.[1]

Perhaps inevitably, Koebele fell in with the officials and growers
who lauded him. His tactless chief, thousands of miles away, could not
compete. Koebele left in the summer of 1891 and returned from

Australia the following spring. He defied Riley's order to ship insects only to USDA agent Daniel Coquillett and sent material also to state bureaucrats. The Californians claimed that shipment directly to them was necessary because Riley and Coquillett wanted the mission to fail.[2]

Koebele searched mostly for predacious ladybeetles. He overlooked the value of the parasitic fly *Cryptochetum* in control of cottony-cushion scale. Nearly everyone did at the time. In the course of his travels over the next two decades, Koebele amassed an extraordinary museum collection of the ladybeetle family, Coccinellidae. His trip to Australia in 1891–92 established several of the predators in California. Riley correctly predicted that they would do little to reduce crop damage, but Cooper and others in the state triumphantly claimed the opposite for twenty years.

Cooper especially touted one ladybeetle as the doom of black scale (*Saissetia oleae*), a citrus and olive pest. He announced that the new imports would finish off the pest within a couple of years. Cooper could see the predators respond to infestations in his own olive orchards. What he did not, and would not, see was that the beetles always appeared in large numbers too late to prevent damaging outbreaks of black scale. Biological control zealots blamed the failure of this predator on some growers' use of sprays and fumigants.[3]

Upon taking office in 1893, new Secretary of Agriculture J. Sterling Morton decided that, if department employees could not operate free from domination by state officials, it was time to pull the agents out of California. Riley and Morton ordered Koebele and Coquillett back to Washington, or else they must resign. Despite layoffs, Riley's abuse, and pay snarls that included a month's salary lost to a Washington bookkeeping error, Coquillett went east and spent his remaining working days as a fly taxonomist. Koebele, however, had already decided to quit. His resignation crossed in the mail with the secretary's ultimatum. Instead of taking Cooper's offer of a regular job as explorer for California, Koebele went to work for the new government of Hawaii at twice his USDA salary. Thus removed from Riley's purview, Koebele promised to control all pests in the islands within three years.[4] Although he could not live up to that pledge,

Koebele remained in Hawaiian employ for the rest of his career and launched that territory's enthusiastic reliance upon biological control.

Riley left the federal service not long after Koebele. Peace with the Californians might have been impossible had Riley remained in his Washington post. He wanted to be secretary of agriculture, or at least assistant secretary in charge of scientific matters. Denied this wish, he made life miserable for those around him until he finally resigned in 1894. California officials absurdly claimed that they had convinced the secretary both to fire Riley and to withdraw Coquillett, whom they called a "kerosene entomologist" for refusing to exaggerate the utility of biological control. Little over a year after resigning, the enthusiastic bicyclist Riley sped down a Washington hill, hit a rock, cracked his skull on the curb, and passed from the ranks of entomologists a few days short of his fifty-second birthday.[5]

Riley's longtime assistant L. O. Howard replaced him as U.S. entomologist and inherited his stormy relationship with California. Howard remained in charge for thirty-three years. Under his direction, the entomological corps grew into a wide-ranging Bureau of Entomology, so named in 1904. Any insect problem in the United States was subject to his attention.

As bureau chief, Howard conceived and actively participated in the most extensive biological control effort of his time—against the gypsy moth—and directed natural enemy work on several other fronts. His own specialty was parasitic wasps, the mainstay of biological control. Later generations of entomologists would consider him a pioneer in the theory behind the method. Yet Howard, more than anyone else, convinced the American public that insects on the whole threatened human civilization, and that civilization must attack them forcefully.[6] He organized his bureau in a way that doomed biological control to dominance by chemicals. The USDA lacked a coordinated team approach and the cultivation of support from industry, which were crucial to the institutional success of biological control in California and Hawaii.

By portraying insects as a dire menace, Howard continued Riley's work of promoting the profession of economic entomology. As Howard became the nation's top employer of entomologists, the profession looked to him for leadership. He easily excelled Riley in relations with superiors and his own staff. In the words of a California entomologist who encountered Howard late in the bureau chief's career, "Unlike Riley, Howard was a loveable character and a smooth, suave, polished gentleman. Where Riley used a battle axe or bludgeon, Howard used a stiletto if the occasion warranted."[7] By acting less like an empire builder than his predecessor, Howard became a virtual emperor of American entomology.

Howard believed that biological control should play an important role, but he warned against expecting too much and sharply criticized others for overstating the benefits of parasites and predators. He tried to restrict states from deliberately introducing insects, for fear that new pests would get in by mistake. By the time he instituted this policy, however, California had firmly established itself in the practice. Howard maintained a semblance of diplomatic relations with the gung-ho Californians in hopes of keeping at least a small measure of control over their actions.

However, first in private and then in public at the end of his career, Howard condemned what others called California's "parasite craze." Haphazard importation of supposed beneficial insects, he wrote, not only risked establishment of new pests from abroad, but set back the development of other methods in applied entomology in California for at least a decade.[8]

The continued development of insecticidal measures by the University of California and by private interests belies Howard's latter claim. University entomologist Charles W. Woodworth, a vehement critic of the State Board of Horticulture, even denied that cottony-cushion scale was under biological control, though few in the citrus belt took him seriously. Ellwood Cooper's own man, quarantine officer Alexander Craw, acknowledged that growers must use chemicals for pests not yet controlled biologically.[9]

Cooper himself accepted only one approach. He professed the belief that God had assigned to each insect a single, specific enemy that would make it too scarce to damage crops. No species was a pest in its native habitat, because "the parasite" prevented it. Cooper seems to have applied the term to both predators and parasites. To control a phytophagous insect, one needed only to determine its place of origin and find the "true parasite" there. What worked for cottony-cushion scale would work for anything else. Whether his subordinates shared Cooper's religious zeal for this approach is not known, but they joined him in claiming victory over pests such as codling moth, black scale, and purple scale, which remained serious headaches for fruit growers many years later.[10]

Some growers suffered crop damage they might have been able to prevent had they not blindly followed Cooper's policy. Better cooperation between the state board and other workers in California could have strengthened entomology in the state. But the "parasite craze" did not entirely retard progress in the field as Howard claimed. Indeed, exaggeration sustained interest in a method that would eventually prove rewarding but that might not otherwise have been given the chance.

Howard remained in correspondence with Cooper and Craw and cooperated with them to have vedalia shipped to other countries plagued by cottony-cushion scale. He even asked for parasites claimed successful against certain pests, so that he might try them in the eastern United States. For several years Cooper and his followers had no money for exploration. California could do little more than ask workers elsewhere to send parasites. This was usually an ineffective way to obtain beneficial insects when so little was known about them and their host relationships around the world. Apparently at Craw's suggestion, Cooper began to demand that other countries pay for any insects California supplied. Howard strongly objected. He finally met Cooper and Craw in person in 1899, on his first of many visits to California. Howard especially liked Craw and hoped that the quarantine officer could restrain the state board's enthusiasm for natural enemies.[11]

Only a lack of money, however, could restrain Cooper. After funds for exploration were restored, he wanted to regain Albert Koebele's services, but could not match his Hawaiian salary. Craw also declined the explorer's job. He and Koebele both recommended Los Angeles County horticultural inspector George Compere.

Born in 1858, Compere managed a citrus grove in what is now downtown Los Angeles by age twenty. He became an inspector in

1891. He admired Craw, who impressed him as a scientific entomologist despite a lack of formal training. Craw at least read some of the literature and moved cautiously. Compere, on the other hand, came to share Ellwood Cooper's unbridled enthusiasm for finding the one "true parasite" of each pest. Harold Compere, who served as foreign explorer decades later, wrote that his father's lack of scientific training made him the perfect choice to pursue Cooper's ignorant policies. Los Angeles County sent George Compere to Hawaii in 1898 to get some of the beneficial insects Koebele had introduced there. The county board of horticultural commissioners hoped to force other counties to buy the insects. In August of 1899, while still in Hawaii, Compere received the appointment to be California's explorer and travel with Koebele.[12]

For the next ten years, Compere roamed the world, seeking enemies of pest insects and perhaps even outdoing Cooper with his boasts of success. After a few months at Koebele's side, Compere took off on his own. He shared Koebele's admiration for ladybeetles at first, but shifted to an equal enthusiasm for parasites.[13] In his zeal for biological control, Compere infuriated entomologists on at least four continents.

Howard's dealings with George Compere reveal the effort to which the chief federal entomologist would go to avoid making enemies. Howard corresponded warmly with Compere, praising his work and gently urging caution in studying, packing, and shipping insects from one country to another.[14] In letters to others, Howard revealed his true feelings:

> Mr. Compere is a man totally devoid of education; a man who has no training in entomology; a man who knows hardly one parasite from another, and who therefore is a dangerous person in this particular quest. He imports all sorts of things into California, and so far has introduced not a single insect which has accomplished any good. His statements are absolutely unreliable, and he claims everything in sight. He is a charlatan.

Howard asked that these words remain private. By maintaining good relations with people he considered unqualified, he could still use them in his own plans for American pest control.[15]

Howard feared that the Californians would, in their ignorance, bring harmful insects to the United States. He worried about hyper-parasites (also called secondary parasites), which fed upon other parasites rather than on the agricultural pests one wished to control. Many parasitic species in the order Hymenoptera develop at the expense of others. Although the adult female of a hyperparasitic species lays its eggs in or on a host such as a scale or caterpillar, the offspring mature by destroying primary parasites. Hyperparasites would presumably reduce the numbers of a primary parasite, perhaps preventing it from controlling the pest population. Someone observing a hyperparasitic wasp emerging from a scale insect might mistake the wasp for a primary parasite.

Careful study would be necessary to determine whether the wasp had fed upon the scale or instead upon another parasite within the scale. Howard believed that he and some of his assistants were qualified to make such judgments, but that the Californians were not. An experienced hymenopterist could search the literature for descriptions of the parasite and its habits. George Compere was traveling the world without a library or a museum collection. Alexander Craw, who received Compere's shipments and reared parasites from them, also lacked the expertise to identify hyperparasites on sight. Craw often sent specimens to Washington to get Howard's opinion. Howard considered him an island of responsibility in a sea of California incompetence, but often warned him not to release any foreign insects without making certain they were not hyperparasites.[16]

Compere and Craw did establish a hyperparasite, *Quaylea whittieri,* in California in 1901. The mistake was not discovered until nearly twenty years later, shortly before the population of the species exploded. For a time entomologists blamed the hyperparasite for the failure of a major biological control project. Howard used the incident to illustrate the danger of allowing biological control efforts to proceed with too much enthusiasm and not enough federal supervision. Although Howard did not know about *Quaylea* at the time he formed his opinions of the Californians, he later treated it as confirmation of what he had believed all along.

Because of the experience he, and Riley before him, had had with California, Howard opposed the efforts of other states to enter the field of beneficial insect introduction. His own bureau made introductions with little more safety than the Californians. As we shall see, Howard himself earned a share of the blame for *Quaylea*. Meanwhile, his restrictions on the activities of other states held back the progress of biological control and helped allow chemicals to predominate.

George Compere sent parasitized black scale (*Saissetia oleae*) to California from Australia in 1900. He declared that control of the pest was at hand. The natural enemy that so excited Compere has never been identified. He called it the "Brisbane parasite." Craw released nothing from the shipments that year, but sent specimens of emerging parasites to Howard for identification. Compere sent another parcel supposedly containing the "Brisbane parasite" in January 1901 and warned Craw to beware of hyperparasites. When several different parasites emerged, Craw particularly suspected one of being a hyper. He sent specimens to Howard and asked directly whether the parasite was primary or secondary. Howard replied that it was a new species, and that an assistant would name it after Craw at Howard's suggestion. He said nothing about the insect's habits. Flattered by the name, Craw took Howard's answer to mean that the species was a primary parasite. Howard probably intended as much, because it fell into a taxonomic group previously unknown to contain hyperparasites.[17]

Craw spread the new parasite throughout southern California under the name *Hemencyrtus crawii,* though this designation was not taxonomically valid because Howard's assistant never published the description. The parasite quickly faded into obscurity. Less specific in its host relations than the insects it attacked, so-called *H. crawii* survived in small numbers among the primary parasites of various soft scale insects. Other entomologists later discovered its hyperparasitic nature and named the species *Quaylea whittieri*. Craw, who died in 1908, never knew of his mistake. A black scale parasite introduced in 1919 at first appeared to spell the end of the pest in some areas, but then *Quaylea* multiplied enormously. The new primary parasite failed to control the scale. The hyperparasite got the blame. *Quaylea* has since virtually disappeared. California entomologists no longer consider it responsible for the failure of biological control.[18]

To Howard's mind, the *Quaylea* incident justified a federal monopoly on biological control. He thought that *Quaylea* had ruined a promising project. He blamed the amateurish, careless work of George Compere. After learning that Compere had warned of hyperparasitism, Howard admitted that he himself might have given Craw the impression that the parasite was primary. Howard still faulted Craw for not making careful laboratory studies to decide the question. Craw trusted him, as the world's leading authority on parasitic Hymenoptera, to predict the habits of species from their taxonomic affiliations.[19] By the 1920s, Howard no longer believed such predictions were reliable, but his correspondence shows that he had believed it when he was dealing with Craw at the turn of the century. Howard's own employees released parasites accidentally on occasion. At least once they deliberately introduced a species that turned out to be a hyperparasite. By luck, it failed to become established. As others in the USDA later acknowledged, only good fortune prevented more mishaps in the early days.[20]

Because he falsely supposed that the USDA could conduct biological control efforts more safely than anyone else, Howard discouraged others from trying. The Californians irritated him in other ways, all of which probably added to his desire to keep states from importing parasites and predators. Howard decried exaggeration, yet was not above it himself. In 1901 he received a Chinese ladybeetle that preyed on the widespread fruit pest known as the San Jose scale (*Quadraspidiotus perniciosus*). He sent the predators all over the country and concluded that they would have controlled the scale, at least in the South, had chemicals not interfered. Actually, the beetles disappeared from America after only a few years.[21] Howard's claim was no more sound than those the Californians made for insects they imported. He saw the two cases differently only because the San Jose scale project involved him and his top assistant.

Howard resented the Californians' stubborn denial of credit to the USDA for the vedalia beetle or for assistance he gave them in arranging for later introductions. Compere's increasing boldness and prominence especially disturbed the federal bureau chief. After Ellwood Cooper ran out of money for foreign travel and called Compere home in 1901, the explorer instead accepted a similar position from agricultural officials of the state of Western Australia. Compere had sold them his promises of success while traveling there. Western Australia hired him chiefly to control Mediterranean fruit fly (*Ceratitis capitata*), a serious pest of numerous fruits in Australia, Africa, Asia, Europe, and South America. (In newly federated Australia, no counterpart to the USDA existed to feud with state officials.) In 1903 the new governor of California rewarded Cooper's support by making him sole commissioner of horticulture and restoring funds for biological control. After this, California and Western Australia shared Compere's services until 1910.[22]

Howard undoubtedly feared for the reputation of all American entomology. Compere bitterly opposed anyone who would disagree with him in matters of credit or reject his claims to have solved a given pest problem. Compere's opponents included entomologists of other Australian states, and especially Charles P. Lounsbury of South Africa. Lounsbury, an American whom Howard had recommended to the Cape Colony government, traveled to Brazil because Compere had reported effective natural enemies of Mediterranean fruit fly there. Lounsbury found nothing controlling the fly. The indomitable Compere insisted that his own failure to establish South American enemies of the medfly in Australia was the fault of assistants in the latter country. Compere later grumbled about incompetence in California as well.[23]

H oward's relationship with California authorities reached its nadir when they tried to interfere with his most ambitious biological control effort, against the gypsy moth (*Porthetria dispar*) in New England. The gypsy moth project was also pivotal in USDA organization of biological control, and in the development of population dynamics theory.

Howard put together a group of entomologists whose sole stated purpose was to attack the pest with natural enemies. On the surface, this mission looked like a federal version of the parasite craze. In

terms of practical success, Howard accomplished no more in New
England than Cooper did in California. In part because this project
failed to defeat the gypsy moth, Howard abandoned the coordinated
approach. He scattered biological control among the various divisions
of the Bureau of Entomology, where too much depended on people
who neither knew nor cared enough about the method. Some of the
most important figures in twentieth-century biological control were
trained in the gypsy moth campaign. One of these, Harry S. Smith,
would take over California's program in 1913. Finally, out of the New
England effort came the first clear articulation of a theory of density-
dependent action by natural enemies. The idea would occupy Smith
and the California school in the years ahead.

As the entomological corps of the USDA grew into a bureau,
Howard followed the suggestions of his assistant chief, Charles L.
Marlatt, and organized it on the basis of crop groups.[24] Occasionally
the chief set up a unit for a particular pest or pests, such as the
Division for Preventing the Spread of Moths. Entomologists working
on natural enemies were assigned to a crop (or insect) division, which
controlled the funds. Resources were generally insufficient to support
all phases of a biological control project, from foreign exploration to
the domestic follow-up work needed to determine the effect of natural
enemy releases in the field. This setup often left essential activities in
the hands of workers who knew little about parasitic insects and had
other things to do, such as testing chemicals. Not surprisingly, the
agent in charge of a field station usually concerned himself with
immediate results that could justify continued funding.

Howard created the moth division specifically to deal with the
gypsy moth, which had been spreading and defoliating hardwoods
since its introduction from Europe in 1869. In 1905, he assembled a
team to receive natural enemies through his own nearly annual visits
to Europe and from a network of corresponding entomologists. At
times, American explorers also took to a field stretching from France
to Japan. Numerous species of parasites attacked the gypsy moth in its
native land. Howard reasoned that all of them might be necessary to
bring the moth under control in this country, although the only

outstanding success in biological control thus far, against cottony-cushion scale, had required just two natural enemies.[25]

Entomologists at the division laboratory in Melrose Highlands, Massachusetts, studied parasites and predators and learned much about packing, shipping, receiving, and rearing insects for release into a new country. The local supervisor of parasite studies, William F. Fiske, speculated as to the sorts of natural enemies explorers should seek first. Fiske defined "superparasitism" as the attack on one host individual by two primary parasites. He concluded that even a combination of species attacking one stage of a host would be unlikely to control it, and that control would require a sequence of parasites of several stages from egg to pupa. (Smith, who worked directly under Fiske at Melrose Highlands, later restricted the term "superparasitism" to parasitism by two members of the same species. Attack by more than one species is now called "multiple parasitism.")[26]

Fiske's idea did not stop the destruction wrought by gypsy moth. The pest still raged after the establishment of about a dozen species of predators and parasites in New England. Howard and Fiske published a book-length review of the project in 1911. One historian has characterized the report as a virtual admission of defeat, or at least an admission that the gypsy moth problem would require a great deal more work. Howard, however, remained confident. Fiske was more of a theoretician, and Howard urged him to minimize the theoretical portion of the report and emphasize the practical side. Yet a discussion of the relationship between population size and the intensity of biotic factors of mortality must be largely Howard's.[27]

A few years before, Howard had taken a tentative step toward a theoretical explanation of how a natural enemy dependent on a pest for survival could keep the pest population under control. After an outbreak of a native species, the white-marked tussock moth (*Hemerocampa leucostigma*), in the Washington area in 1895, he observed that the population of parasites rose sharply. The host nearly disappeared by the end of the season. In general, Howard supposed, the rise of a host would lead to such rapid multiplication of the parasite that the latter would overtake the host in numbers and nearly exterminate it. The outbreaks themselves he considered deviations from normal balance, but he did not attempt to explain their origin. Although Howard clearly believed that parasitic attack increased with the host population, he did not specify that the proportion of parasitism would fall as the host population decreased. Neither did he in any way connect his first theoretical foray with the practical problems of introducing natural enemies to new locations.[28]

In the course of gypsy moth work, Howard and Fiske enlarged upon Howard's earlier discussion. The authors assumed, as others did, that "normal" conditions kept a population at a relatively stable, equilibrium level. To maintain this stability, Howard and Fiske argued, some cause of death must take a greater proportion when the insect was abundant than when it was rare. Such an agency they called "facultative." It would relax when the population shrank, and intensify when the population grew. Since they used the word "abundance" without defining it, it is possible that Howard and Fiske meant raw population size rather than number per unit area. The phrase "population density" was not yet in use. The authors seem to have meant abundance over an undefined area; if so, then their "facultative agencies" were what would later be called "density-dependent." Most mortality factors, Howard and Fiske wrote, acted independently of population size and would therefore not be capable of stabilizing it. These were called "catastrophic." Parasites were the most likely "facultative" agency because they multiplied when their food supply increased, whereas climatic factors would not be affected.[29]

Logically, a factor that changed intensity in this way could maintain the pest population at some equilibrium level, or at least within relatively narrow limits. Entomologists of the later California school of biological control would adopt density-dependent action as the theoretical basis for their work. Howard and Fiske did not mathematize the concepts beyond their qualitative definitions. A relatively steady state was the "normal" situation, which the gypsy moth parasite project aimed to restore. Howard's earlier analysis had concerned outbreaks, which both authors still considered exceptional. The new discussion of density-dependent factors covered the rule, namely stability.

Howard and Fiske parted company after writing the gypsy moth report. Howard continued to stress restoration of parasites the moth had left behind in Europe, while Fiske emphasized other aspects of the environment, particularly the complex of trees. Fiske no longer believed that parasites alone would restore the balance. He concluded that the gypsy moth could be controlled only by a conversion to mixed forests containing few of the especially vulnerable oaks. This idea originated in the U.S. Forest Service, another USDA bureau working on the same problem. Fiske adopted the Forest Service viewpoint after studying the relationship between the gypsy moth and forest types in Europe.[30]

Howard and Fiske wrote an entirely verbal discussion of mortality factors. It formed only a brief section of their lengthy tome. In his separate paper on superparasitism, Fiske began to mathematize host-parasite interactions. He graphed the percentage of hosts attacked in laboratory experiments as a function of the number of eggs laid, including the extras laid in already parasitized hosts. He produced a curve asymptotic to the 100 percent line. The graph matched what would be expected if chance alone determined host selection. Fiske concluded that ovipositing females did not discriminate between parasitized and unparasitized hosts. Only a parasite population many times the size of the host population could be expected to find every last host individual.[31]

This result contradicted the assumption of absolute discrimination upon which French entomologist Paul Marchal had constructed a model for pest outbreaks and their suppression by parasites. Marchal, too, believed that outbreaks were the exception, balance the rule. He assumed that host and parasite populations generally existed at equilibrium, but not necessarily with each other. A population below this level would increase rapidly, slowing as it approached equilibrium. Although he lacked an algebraic expression for this phenomenon, Marchal's description qualitatively resembled the logistic curve developed by Belgian mathematician Pierre-François Verhulst a half century earlier. Marchal argued that, if their host reached equilibrium before they did, parasites would continue to multiply and thus overwhelm the host population and drive it nearly to extinction.[32]

Marchal used a simple numerical example. Even if the parasite did not reproduce faster than the host, he argued, the parasite's increase would take an increasing percentage of hosts. When the host population was at its highest, all of its offspring would be parasitized and there would be no next generation. Marchal's figures show that he assumed all parasite eggs would be laid in hosts, and none would be wasted in superparasitism. He acknowledged that crashes to extinction did not generally happen in nature. Other factors such as competitors, generalist predators, hyperparasites, and weather tended to smooth out fluctuations and maintain the balance he and others usually perceived.[33] Marchal did not, however, analyze the equilibrium situation he considered normal. Howard and Fiske did. Later theorists would build on the work of all three.

Although Howard and Fiske's delineation of density-dependent and density-independent factors was cited as the first until the 1960s, it was actually anticipated by, of all people, Charles W. Woodworth of the University of California. Howard and Woodworth each apparently arrived at the concept of density dependence without knowledge of the other's work. Woodworth's version clearly applied to density—number per unit area—but went unnoticed for fifty years and influenced no one.

As a critic of biological control, one who even doubted its efficacy against cottony-cushion scale, Woodworth was hardly a man to whom biological control advocates would look for ideas. Woodworth believed that the deliberate addition of a previously absent species of natural enemy would not usually alter the target population. In 1908, he characterized natural enemies as density dependent and climatic factors as density independent. He did not give the concepts names; if he had done so, his paper might have been remembered. Woodworth argued that the response of most density-dependent factors would be limited and would not overtake the host. Only in rare cases would a parasite increase enough to stop the increase of its host, and only these rare parasites would be worth importing for biological control.[34]

Howard's (or Woodworth's) mechanism for population balance engendered little comment before the late 1920s. Entomologists who began to challenge Howard's view at that time claimed that it had been almost universally accepted without verification.[35]

The notion of balance had indeed been widely accepted even before Howard and Fiske wrote, but without discussion of density-dependent action. During the same period, Frederic E. Clements articulated a strong position of balanced communities determined by climate and achieved through a succession of stages until a "climax" was reached. His early works dealt primarily with plant ecology in undisturbed environments. Ecologist and entomologist Victor E. Shelford developed a similar model for animal communities in 1913 and briefly described the way natural enemies would end a pest outbreak. Following the increase of a pest, its natural enemies would increase, cut the pest population down, and then decline in numbers themselves. Shelford did not discuss density-dependent action of one species on another to maintain equilibrium.

Shelford emphasized the ripple of adjustments across every species in the community. Economic entomologists such as Howard usually studied the action of a single species on another. They focused on one pest of one plant at a time in an artificially modified system. To the extent that they were ecologists, they were autecologists who perceived little common ground with synecologists. Even after Howard and Fiske wrote their report, neither entomologists nor ecologists gave much thought to density dependence until the late 1920s. They did not applaud or debate it. Neither Howard nor Woodworth ever returned to his theoretical speculations on insect populations.[36]

California horticultural officials at the turn of the century needed no more theory than Ellwood Cooper's assertion that God had appointed "the parasite" to control each insect. To them, anything Howard said was suspect, and anything Woodworth said was wrong. Cooper denied Howard's premise that control of the gypsy moth might require a large complex of parasites. Cooper claimed, as always, that there was one "true parasite." He promised the state of Massachusetts that he could control gypsy moth quickly and completely for $25,000. He offered not only George Compere's services but also those of Albert Koebele, although Koebele was not his to offer. Cooper also offered to find "the parasite" of the boll weevil for southern states.[37]

Cooper's offer to Massachusetts reflected his own shortage of funds. California had trouble keeping its best people because it could not match the salaries offered by the sugar planters in Hawaii. Koebele was only the first to go. Alexander Craw left in 1904 to become Hawaii's plant quarantine inspector at twice his old salary. His California duties were divided between a new quarantine inspector and a man in charge of handling beneficial insects shipped from abroad. The inspector, an entomologist Howard regarded highly, took the Hawaiian position four years later upon Craw's death.[38]

Craw's successor in biological control in California was Edward K. Carnes, a young political disciple of Commissioner Cooper. Other than his espousal of Cooper's philosophy, Carnes's only qualification for the position was a year of experience assisting Craw in San Francisco. Carnes oversaw the expansion of the receiving station from a room in San Francisco to a new facility built for the purpose in Sacramento. The 1906 earthquake hastened the transfer by damaging the laboratory. Carnes moved what was left to a makeshift greenhouse in an assistant's backyard in Sacramento until the new building was completed in Capitol Park in 1908.[39] By the time the State Insectary was ready, however, California had a new state commissioner of horticulture.

Howard had had enough of Cooper's interference with other states and with his own department. The twists of state politics finally brought Cooper down in 1907. Corruption filled much of California state and local government during this period. The Southern Pacific Railroad largely ruled the state until the Progressives came to power in 1911. The political quagmire included the horticultural commissioners, although accounts of the Progressive movement in California do not mention them and barely mention agriculture at all. Having allied with one governor, Cooper found his job in jeopardy when the Southern Pacific machine installed another, James N. Gillett, in 1907.

Howard and other enemies of Cooper prodded the new governor for months.

Governor Gillett finally replaced the fanatical Cooper with John W. Jeffrey, a Los Angeles County horticultural commissioner. Jeffrey won Howard's crucial endorsement by promising to be more realistic about biological control, use other methods such as fumigation when necessary, keep a tight rein on George Compere, and send specimens to Washington for identification before releasing any supposedly beneficial insects.[40]

Howard nevertheless found Jeffrey's administration nearly as troublesome as Cooper's. Only by repeated pleading could Howard persuade the new commissioner to stop Compere from sending poorly packed moth material to Washington, which lacked adequate insect quarantine facilities. Howard wanted Compere's shipments to California stopped as well, but Jeffrey kept the explorer on the job. Jeffrey had the State Insectary built according to the recommendations of Compere and Edward Carnes. Howard visited in 1908 and admired the facility, but feared Carnes's lack of expertise.[41]

Like others in those corrupt days in state government, Carnes worked mainly to improve his own position. What he lacked in training he made up in ambition. To attract public support, he tried to turn the Insectary into an entomological museum, which would indeed have bolstered the facility's intended use had Carnes known much about insects. He failed to establish any beneficial species from Compere's shipments or by correspondence with foreign entomologists. As Craw had done, Carnes released a hyperparasite after consulting Howard. Just as before, Howard relied on taxonomic affiliation to separate primary from secondary parasites and failed to recognize this species' hyperparasitic status. He named the parasite for Carnes, who, like Craw years earlier, was delighted.[42]

George Compere, in all his travels, failed to establish even one beneficial insect species in California, either before or after Carnes took over the domestic end of the work. Only the hyperparasite *Quaylea whittieri* survived, and that was not Compere's fault. He accused workers in Western Australia and California of mishandling his shipments, although Carnes at least appears to have done all he could to release those parasites that contemporary authorities believed were both new and beneficial. Compere resigned his Australian position and refused to explore for California as long as Carnes remained on the job. After the discovery of Mediterranean fruit fly in Hawaii, California strengthened its plant quarantine service. Commissioner Jeffrey named Compere chief quarantine inspector at San Francisco, where he spent the rest of his working days.[43]

Compere achieved neither immediate practical benefit nor advancement of entomological science. His work may have had one disastrous consequence of the sort that L. O. Howard feared most, although Howard seems never to have suspected it. In 1909, Compere sent parasitized Mediterranean fruit flies to Hawaii, where entomologists hoped to control a related species. He warned against letting any healthy flies escape. A year later the medfly was established in Hawaii, and Compere called for stiff quarantine measures to keep it out of California. The fly was first found in the Honolulu insectary where Compere sent his parasites. Entomologists have assumed that the medfly flew to land from an infested shipload of Australian fruit. This is possible, but it is also possible that primitive biological control work brought the species to Hawaii, where it prevented the production of many fruits for export. A University of Hawaii entomologist suggested that Compere's shipment introduced the pest, but this suspicion was allowed to fade quietly.[44]

As Compere's travels ceased, Jeffrey and Carnes emphasized the distribution of parasites and predators native to California. Chief among these were a parasite against certain soft scale insects in stone fruits, and the common ladybeetle *Hippodamia convergens* against various aphids. Both beneficial insects were popular with growers. Carnes rejected arguments, later to become consensus, that the species were already so numerous that what the State Insectary added to the field could make no difference.

Carnes especially touted the *Hippodamia* program. The species hibernates in extremely large aggregations in the Sierra Nevada over the winter. Carnes and his assistants would scout for aggregating masses in late autumn and return a few months later to dig them out of the snow. Shipped by the hundreds of pounds to melon growers in the Imperial Valley of extreme southern California, the beetles were released as aphid populations rose in the spring. In 1912, Carnes's last year on the job, the State Insectary distributed an estimated 43 million *Hippodamia,* free of charge. The program, centerpiece of California biological control by that time, was so popular that Carnes could not keep up with all of the requests.[45]

Like the vedalia, *Hippodamia convergens* was a conspicuous, brightly colored ladybeetle. This trait doubtlessly contributed to its popular appeal. Melon and stone fruit growers added their voices to those of citrus growers in support of biological control. Most of the introductions Koebele and then Compere attempted between 1888 and 1910 were directed at citrus pests. Commissioner of Horticulture Ellwood Cooper cultivated the support of the citrus industry. His successor, John Jeffrey, emerged from a citrus county when Cooper's political base eroded. Citrus growers, beset with several foreign scale insects, hoped for a miracle to match that against cottony-cushion scale. The growers listened to inflated promises of parasitic control. Although he may not have shared Cooper's zealous devotion to biological control, Jeffrey continued efforts in this direction because growers wanted them. Propaganda kept biological control alive when the actual results could not have justified it.

Outside of California, biological control was permitted to grow freely only in Hawaii, which had lured explorer Albert Koebele away in 1893. Interest in the "Koebele method," as it was sometimes called, continued to run high after the United States annexed the islands in 1898. In Hawaii as in California, one powerful agricultural industry came to support biological control and was often at odds with federal authorities. Hawaii's position as the hub of Pacific commerce opened the way for species long blocked by geographic isolation. The territorial government enacted plant quarantine laws and sought beneficial insects to protect industries such as cattle and sugar, themselves foreign to the islands.

There was no pretense of democratic appeal in Hawaii, where sugar dominated the economy. A small number of plantation owners, down to about fifty by the turn of the century, controlled virtually all cane in the islands. The oligopoly formed the Hawaiian Sugar

Planters' Association in 1895, primarily to maintain ironclad control over labor. After alien insects began to attack sugar cane, the association started its own experiment station, creating a division of entomology in 1904. The planters paid half of the handsome salaries of territorial entomologists, and hired others, to ensure that biological control in the islands would focus on sugar. The HSPA financed the Hawaiian Entomological Society and hosted its meetings.[46]

The lords of the sugar plantations took action themselves because the USDA directly opposed them. Universities operated agricultural experiment stations in the states, but the USDA itself ran territorial stations, including the one established in Hawaii in 1902. Secretary of Agriculture James Wilson hoped to fill the islands with white Americans on small family farms and undermine the power structure of the sugar planters, who relied on native or Oriental labor. Wilson insisted that the station staff, including entomologists, work on crops other than sugar.[47]

The planters' own entomologists pressed on. They put their faith in natural enemies, a good bet because nearly every pest was foreign. Even if there had existed insecticides effective against the pests, application in dense cane fields was virtually impossible in the days before aerial spraying. After Koebele retired, other explorers carried on the cause. One nearly died, first from typhoid and then from malaria, in a three-year campaign to bring natural enemies of a sugarcane weevil to Hawaii from obscure South Pacific islands without direct shipping connections. He got the prize home. By the 1950s the Hawaiians claimed to have controlled ten of their eleven sugar pests biologically.[48]

The Territorial Board of Agriculture and Forestry handled biological control on other crops, especially after the only entomologist remaining at the USDA experiment station defected to the board. Pineapple insects received increasing attention as that fruit became a major export crop. The USDA planned to use pineapples to break the grip of sugar plantations, but the plantation system monopolized the new crop as well. An industry-sponsored Pineapple Research Institute, founded on the University of Hawaii campus in the 1920s, supported exploration for parasites and predators. Biological control was less successful in the pineapple industry than in sugar, so growers had to rely more on chemicals. However, exploration for natural enemies remained the first line of attack against any new pest that invaded the islands.[49]

Hawaii continued to support biological control because those in authority, bolstered by powerful private interests, believed the method to be worth continuing. It did not matter how misguided the federal authorities and mainland entomologists thought this enthusiasm might be, as long as the sugar and pineapple people were willing to pay for the work. A team of entomologists dedicated to using natural enemies handled all phases from exploration to field evaluation, while appreciative farmers cheered. The group remained rather isolated from mainland entomology. Success in Hawaii could easily be dismissed either as propaganda or as something exceptional, effective perhaps in tropical island conditions but not for general application to American agriculture.

Institutional factors similar to those in Hawaii sustained biological control in California, but workers in the latter place strove to be part of mainstream American entomology. Unlike Hawaii, California also built a large insecticide program. California ultimately produced more significant scientific developments in biological control, trained more workers, and constructed a more compelling case for including natural enemies in the overall pest control picture.

Neither California nor Hawaii could prove much practical success early in the twentieth century. Reaction against the swaggering Californians stifled efforts in other states. L. O. Howard feared that exaggeration of biological control would delay the development of other methods and damage the reputation of his profession. In the long run, however, overuse of chemicals had that effect. The choice of gypsy moth for a concerted biological control program in 1905–11 was unfortunate, because its failure led to the scattering of USDA biological control efforts to the point that none had much chance of success. Parasite introduction against the moth continued until 1933 and resumed in the 1960s, but never achieved control.[50]

Howard still believed that the gypsy moth project could succeed, but he would no longer commit all of his top parasite specialists to it.

He reassigned them to various divisions in the Bureau of Entomology. Usually only the explorer would be dedicated to biological control. In some cases, another agent receiving shipments from abroad might share this dedication. USDA specialists in parasite systematics, however, worked in Washington, far from the farmers' fields. Field entomologists, who worked closest to the farmers and who were under pressure to produce immediate results, had neither the time nor the inclination to implement biological control and evaluate its effects.

In light of his experience with the Californians, Howard did not wish for other states to import parasites and predators. Only a few states expressed interest, anyway, and he discouraged those. He believed he had the power to stop the state of California and the territory of Hawaii from making importations. Instead of testing whether he really had such legal authority, he used it as a threat to try to keep the enthusiasts in line.

Biological control especially interested officials in Florida, another citrus state, but they lacked the means to match the efforts of the Californians. In many respects, the development of the Florida citrus industry paralleled that in California, in nearly the identical time period. However, California's chief competitor in citrus never formed a marketing organization with the clout of Sunkist. The fact that Florida growers did not match Californians' early success in lobbying for state pest control aid supports historian Howard Seftel's contention that such lobbying paved the way for cooperative marketing. Direct comparison is difficult, however, because a devastating 1894–95 freeze set the Florida industry back several decades. From 1935 onward, the state itself conducted intensive advertising for Florida citrus.[51]

Florida moved into biological control in a small way quite early. Beginning in the 1890s, the state distributed endemic fungi for control of scale insects and whiteflies. For the next half century, Florida entomologists defended the practice, despite a lack of evidence that it improved on naturally occurring conditions, and despite a U.S. Bureau of Entomology study strongly suggesting that it did not. The State Plant Board even offered fungus cultures for sale. Some Floridians, however, wanted to do more.

Howard dispatched Russell S. Woglum (later entomologist of the California Fruit Growers Exchange) to Asia in 1910, rather than let Florida interests import natural enemies of the citrus whitefly (*Dialeurodes citri*) themselves. Woglum found a promising species (*Encarsia lahorensis*), but it failed to survive in laboratory culture in Florida, and no further attempt was made to import the parasite until fifty years later.[52]

Howard sent other searchers out in the 1920s specifically to head off states from joining the field of foreign parasite introduction. Oregon alone on one occasion and a group of northwestern states on another were ready to sponsor explorers. Howard argued that only his bureau had personnel qualified for such work, which would be dangerous in other hands, although by that time he had made peace with California.[53]

Howard did not intend that his administrative setup should de-emphasize biological control, but that was the result. The USDA had access to superior insect collections, libraries, and assistance in foreign countries. Biological control would usually not be as easy as it had been against cottony-cushion scale. Substantial, persistent effort would be required to determine which pests could be beaten this way. Howard had the most entomologists and was in the best position to put together a team that could really make the method work. Success would require a specialized unit dedicated to biological control rather than to whatever method could kill a few insects quickly and conspicuously. Biological control offered long-term solutions but required long-term research, whereas Howard organized his bureau in a way that forced short-term goals to the fore.

California, on the other hand, did commit a team to biological control. This team, prior to 1913, did not acknowledge the difficulty of the method. Ellwood Cooper's band of enthusiasts lacked the sort of entomological training Howard would have preferred. The Californians' dedication was politically motivated. They, like the USDA, failed to control any insects biologically in the quarter century that followed the cottony-cushion scale episode. Only boasting and political maneuvering preserved a program whose sole objective was the control of insects by natural enemies. To the legacy of dedication from the "parasite craze," later generations of scientists brought a more realistic approach and a broader knowledge of insects. The success of a chemical treatment in this or that case did not displace biological control from the research agenda in California, although it did in the federal government.

Biological Control
Comes of Age

Biological control produced more words than practical results in California during the early part of the twentieth century. After the arrival of Harry Scott Smith in 1913, the program evolved into something that scientists could respect. Eventually this would help the citrus growers whose support had kept biological control alive. Smith moved quickly to curtail exaggeration and to heal rifts with the U.S. Department of Agriculture and the University of California.

After ten years in Sacramento, Smith moved to the university and gradually pursued a more rigorous program of research in addition to agricultural service, although the latter always remained his primary purpose. Just as he and his staff found some real success in the citrus groves in the late 1920s, Smith began to delve into theoretical issues. He joined ecologists in debating population dynamics. Practical problems and experiences in biological control of orchard pests determined the scientific questions—and many of the answers.

Two decades after the spectacularly successful control of cottony-cushion scale, the state of California still maintained an official agency dedicated to the biological method. Under the auspices of the Commission of Horticulture, the State Insectary moved predacious and parasitic insects into and around California. Unfounded boasts still marked the activity through 1912, however. Certain groups of farmers eagerly supported biological control, but were bound to realize sooner or later that the program delivered a good deal less than promised. Without their endorsement, a unit whose existence depended on politics could collapse.

Dissension in the ranks crippled much of the activity of the Insectary prior to 1913. Foreign searches for new beneficial insects stopped, largely because the explorer could no longer get along with the man receiving his shipments. Entomological authorities in the USDA considered California's exploration dangerous and were glad that it had come to a halt. The federal government strengthened its own regulations concerning the movement of insects and passed the first national plant quarantine law in 1912. USDA entomologists may have acquired the authority to shut the California work down should it begin again, but their own efforts in biological control were organizationally crippled.

The coming of Smith set in motion a series of changes that secured the future of biological control in California. He brought recognized scientific training and experience with parasitic insects. Most important for the future, he came from the USDA and brought the cautious attitude of that agency into the State Insectary. With his ties to the federal department, he made peace between Washington and Sacramento, so that the USDA did not try to stop him from resuming importation of foreign insects. Smith also adopted the economic interests of farmers, especially the specialized citrus growers of southern California, with whom he established personal contact. He accepted their goals of high productivity, cosmetic perfection, and maximum profit.

Employment in state government restricted the freedom for which Smith yearned as a scientist. He had always wanted an academic position, which he attained when the entire biological control program was transferred to the University of California in 1923. Without compromising his commitment to the profitability of California agriculture, Smith cultivated a new audience of other scientists. The university setting encouraged Smith to delve into theoretical population ecology, which became a major focus of his later career as it could not have been during his years of government service. When he addressed theoretical concerns, however, he did so from his agricultural perspective. Theory came as a response to practical challenges and experience, although the ideas thus generated rarely altered the practice of biological control. For decades, other scientists fiercely debated the ideas to which Smith had contributed, but he generally stood clear of the fray.

Smith's balancing act between public servant and pure scientist illustrates the difficulty of establishing agricultural scientists' place in the relationship between modern science and technology. Over the last thirty years, historians of technology have sought to counter the view that information flowed only one way, from basic science to application. According to a model advanced especially by Edwin T. Layton since 1971, science and engineering are distinct but equivalent communities, each with its own body of knowledge and building mostly on itself. The relationship between science and technology, which Layton called "mirror-image twins," became reciprocal rather than hierarchical.

Historians have used the interactive model to examine technological knowledge in its own right and to consider the influence of technology on science. Various case studies, including examples from medical science, have shown that the path from pure science to technology can often be reversed. However, the model of two distinct communities was developed around physics and corresponding engineering fields.[1] Even if workers in those areas can be identified as either scientists *or* engineers, agricultural scientists such as the entomologists in Harry Smith's group are not so easily labeled. Agricultural experiment stations indeed represent a "peculiar juncture of pure and applied science."[2]

This chapter explores the relationship between pure and applied science with respect to Smith's work on population ecology from the late 1920s into the 1930s. Chapter 5 does the same for his and his associates' approach to systematics and evolution in the 1930s. In both cases, the pure science grew out of specific technological problems in pest control.

Complete practical successes remained few. During Smith's first twenty years in California, he and his staff achieved total control of only one more pest by natural enemies. Nevertheless, the method survived and gained an aura of scientific respectability that had been lacking. Work with beneficial insects retained a degree of institutional autonomy found neither in the federal government nor in any other state. With this combination of dedication and realism, biological control came of age.

By 1911 the office of California State Commissioner of Horticulture John W. Jeffrey was a mess. Jeffrey's subordinates, described as a "gang of fighting narrow-minded puppets," were at each other's throats. The Progressive campaign had just swept the Southern Pacific Railroad regime from power in Sacramento in the 1910 elections. Advisers urged new Governor Hiram W. Johnson to remove Jeffrey as representative of the old guard.[3]

Johnson deliberated for nearly a year before replacing Jeffrey with Pomona College biology professor Albert J. Cook. The horticultural commissioner dealt primarily with insect problems, and in Cook the state at last could boast of an eminent entomologist in the position. He was among the pioneer teachers of entomology, beginning at Michigan Agricultural College (later Michigan State University) in 1867. After moving to Pomona in 1894, Cook helped improve relations between citrus growers and the University of California in the days before the university established a cooperative extension service or had any permanent southern campus.[4] Having assisted the university in this way, he was able to end the old feud between Berkeley and the horticultural bureaucracy in Sacramento.

The aging entomologist had become an important voice in southern California agricultural politics by 1911. Several of his students were moving up in the ranks. One of these, the Ventura County horticultural commissioner, developed ties to the most influential man in the citrus industry, Charles C. Teague. Teague's endorsement helped persuade the governor to choose Cook for the state office. Two years later, former commissioner Jeffrey tried to discredit Cook and have him removed, but an investigative team led by Teague convinced the governor to retain Cook in office.[5]

To the federal government, Cook's appointment signaled a possible end to the era of carelessness and exaggeration in biological control. He had once collected moth parasites for the USDA while traveling in Europe, but he had also been among the earliest advocates of chemical treatments such as kerosene emulsions and arsenical dusts. U.S. Bureau of Entomology Chief Leland O. Howard considered Cook someone with whom he could work calmly and profitably, for Cook was one of the grand old men of the entomological profession. Years later, Howard looked back at Cook's selection as having brought California out of its "fog of parasitic control."[6]

The new commissioner strengthened the state's plant quarantine service and helped reshape the state and county horticultural commissions into a professional civil service under the Progressive administration. As a part of a reform government, with his own ties to university science and to leading farmers, Cook could settle feuds and build on what was valuable in the legacy of his predecessors. He took over a service that already included a unit devoted to biological control. To run it, he hired one of L. O. Howard's favorite young scientists, one who maintained proper deference to federal authority and, in particular, to Howard himself. This relationship allowed biological control to mature in California as in no other state.

At first, Cook was content to retain most of his predecessor's staff, including State Insectary superintendent Edward K. Carnes. The state no longer employed an explorer in foreign countries. Carnes was distributing beneficial insects already in California, especially the native ladybeetle *Hippodamia convergens* against aphids and the parasite *Encyrtus californicus* (then known as *Comys fusca*) against certain scale insects. He wrote to entomologists around the world to try to set up insect exchange through the mail. Although several responded, Carnes did not establish any new beneficial species in California as a result.[7]

After nearly a year, Cook asked George Compere if he would like to go parasite hunting again, but Compere refused to explore as long as Carnes remained at the Insectary. The old explorer accused

Carnes of mishandling shipments and stealing insect collections belonging to the state. The dispute over collections marked the final break in an old friendship. Cook and his staff, stunned by the charges against Carnes, investigated his work more closely and were surprised to learn that all beneficial species established in California had come before his time. Accepting the judgments of entomologists far and near, the commissioner found Carnes incompetent and forced him to resign. However, like others before him, Carnes eventually found a job with a high salary in Hawaii.[8]

To run the State Insectary, Cook wanted a trained entomologist with expertise in parasitic insects. Instead of choosing a political ally, as previous commissioners had done, he went to the USDA and recruited Harry Scott Smith, one of the most promising young agents in the Bureau of Entomology.

Born in 1883 to a poor Nebraska farm family, Smith earned bachelor's and master's degrees in entomology at the University of Nebraska under Lawrence Bruner. Bruner trained perhaps more prominent entomologists than anyone else in the period save John Henry Comstock of Cornell. Smith intended to pursue a career in business and distance himself from the poverty of his childhood, but his success in an entomology course led Bruner to hire him as a laboratory instructor. The professor opened his home to his students, and none gained more from the invitations than Smith, who could rarely afford a good meal. Ultimately he married Bruner's daughter Psyche, who was apparently so named at the suggestion of L. O. Howard. (Besides the character from classical mythology, the name represented a family of moths and an entomological journal.) Psyche Smith hated her given name and would have nothing to do with entomology herself, although her father, her husband, and two of her children were entomologists.[9]

Lawrence Bruner specialized in grasshoppers. His only connection to biological control was an unsuccessful collaboration with Howard to use a fungus against grasshoppers.[10] As a student, Smith focused on the order Hymenoptera, doing his graduate research on a group of sand-dwelling wasps that feed on grasshoppers. He moved into parasitic forms after joining the Bureau of Entomology in 1908.

He worked on boll weevil parasites in the cotton belt for a few months. The bureau, which moved people around often, transferred Smith to its largest biological control project, on the gypsy and brown-tail moths in New England. The young Nebraskan spent a year and a half at the Melrose Highlands, Massachusetts, laboratory and became the most favored assistant of supervisor William F. Fiske.[11]

Smith made his early reputation especially with the discovery that a native hyperparasite underwent hypermetamorphosis—that is, passed through two very different larval forms. The species had penetrated a lab considered to be insect-tight, without the primary parasite on which the hyper must feed. Its unusual first-stage larva, which had defied identification, burrowed into the host caterpillar and awaited the arrival of a primary parasite. Most parasites, whether primary or secondary, were attached directly to the host as eggs, but these strange larvae moved freely outside their hosts before finding their way in. Smith was unable, until several years later, to verify his suspicion that the adult hyperparasite laid eggs on foliage, leaving the larvae to make contact with host insects. The upshot for biological control was that entomologists needed to be even more vigilant against the introduction of hyperparasites than was previously thought.[12]

Smith hoped to get a job at a university and agricultural experiment station in the West. No such position came his way, so the restless young man took a leave of absence to start a homestead in Wyoming, where he had been looking to invest in irrigated ranch land. The venture collapsed, and in a year Smith was back in Melrose Highlands. The Bureau of Entomology soon transferred him again in the scattering of the gypsy moth staff.[13] Assigned to biological control of the alfalfa weevil (*Hypera postica*), a European insect that had been introduced to Utah, Smith moved from laboratory study to the more practical side of parasite work. He spent a season in southern Europe searching for parasites of the weevil, then moved to Utah and took charge of those he and others had sent.

The way the USDA organized biological control after breaking up the gypsy moth group handicapped work on the alfalfa weevil. The man in charge of the lab near Salt Lake City knew and cared little about parasites, yet had responsibility for releasing them and following their progress until Smith arrived from Europe. Smith complained about the handling of his shipments and about the scant attention paid to biological control in the Division of Cereal and Forage Crop Insects. Only one parasite, *Bathyplectes curculionis,* was established. Although it later proved more effective in other areas to which the weevil spread, *B. curculionis* did little to reduce alfalfa damage in Utah. Eager to move on, Smith found an opportunity after he had been in Utah for less than a year.[14]

It is sometimes mistakenly believed that L. O. Howard recommended Smith for the California position. Horticultural Commissioner Cook asked Howard if Smith was available. The federal bureau chief reluctantly replied that he would not block a good career move. Howard urged Smith to decline the job and avoid entanglement in California politics. Smith's division leader accused Cook of crippling the alfalfa weevil work and thus endangering even California, to which the pest might (and eventually did) spread. Promotions came slowly in the Bureau of Entomology, where Smith made only $1,700 per year. Cook offered not only good facilities and full charge of revamping the state program, but also a salary of $2,400. Smith accepted, and took up his new post in Sacramento at the beginning of 1913.[15] Cook lived only three more years, but his hiring of Harry Smith completed the legacy state politics left to biological control.

Smith immediately hired an assistant from the USDA alfalfa weevil project. This further dismayed Howard, who complained that hiring people whose superiors wanted to keep them constituted a breach of ethics. Smith soon soothed the bureau chief's anger. He convinced Howard that, from now on in California, biological control would proceed with caution and follow methods the USDA used at its gypsy moth lab. Smith promised to send specimens to Washington for approval before releasing any supposedly beneficial species.[16]

Most important, Smith tempered years of exaggeration of biological control in California and went out of his way to avoid grand public predictions. Over the years, Smith's views came closely to parallel Howard's. Both men advocated a limited role for biological control, warning that it would never answer all insect problems and that those it could solve would take time and careful research. Smith conceded that his predecessors in California had made rash promises and had risked the introduction of new pests. Smith specifically rejected biological control for insects such as the boll weevil and Colorado potato beetle, which had become pests due to manmade changes in their native environments rather than by sudden introduction from another part of the world. Presumably no effective natural enemies would be found for such an insect because, if they existed, it would never have become a pest.[17]

Smith did not suggest that farmers change agricultural practices to make crops less hospitable to pests. It is not clear whether he ever considered this a viable method of pest control. His brief experience with the boll weevil in the USDA may have convinced him that farmers would never accept complex cultural measures—and he always kept the farmers in mind.

Howard gave Smith wholehearted approval and later declared that it was only because of Smith that the USDA did not stop California from importing insects. The two men deeply admired each other, and they cooperated by informal agreement until Howard retired. For years, they exchanged data for a parasite host catalogue and biological control manual that they planned to write but never completed. And when Smith wanted to hire someone from the federal bureau, he asked Howard's permission first.[18]

California no longer had to fight federal authorities to keep its own biological control program going. In the future, the state could freely share in the work of USDA explorers and taxonomists, as well as make sure farmers gave natural enemies a real chance in the fields and orchards. The USDA rarely did the latter. The Bureau of Entomology sought quick fixes to insect problems, relegating to low priority a method that might take a long time to implement. The California State Insectary and its university successor always sought biological solutions, because the agency existed only for that purpose. Under Smith, the chances of actually finding those solutions improved.

To halt the propagandistic practices of the past, Smith had to terminate several of the most conspicuous projects of the State Insectary, despite their popularity. In 1914, he stopped colonizing a highly touted parasite George Compere had introduced to fight codling moth. Repeated releases of large numbers had failed to establish the parasite, which Smith called useless. He also ceased distributing a native parasite that had been a favorite project of Edward Carnes. Smith found it difficult to convince stone fruit growers that artificial distribution served no purpose. "You have more *Comys fusca* on one tree than we could possibly send during the entire season," he told them, but requests kept coming in.[19]

Political necessity forced Smith to continue sending out colonies of vedalia beetles on request. Those already in the field nearly always caught up with isolated outbreaks of cottony-cushion scale before the shipments reached the growers. In the lab, the beetles overran their food supply so quickly that the staff could not keep them in stock. Rare infestations of the scale had to be found to start new colonies of vedalia. The best places to find cottony-cushion scale were in areas treated with chemicals to fight other insects. For example, in one of the earliest clear cases of a pest outbreak caused by sprays, lead arsenate killed the vedalia beetles in Santa Clara County pear orchards in 1915.[20]

Most prominent of all Insectary activities at the time Smith took over were the annual releases of millions of *Hippodamia convergens* in the Imperial Valley. Known as the convergent ladybeetle because of an angled pair of white spots on the back of the thorax, the species could just as easily have taken its name from its huge winter aggregations. These were collected in the mountains and shipped by the ton to cantaloupe growers. Smith cautioned against expecting benefits from the releases, which reached 100 million ladybugs per year. Until he could prove them useless, however, political pressure required him to continue the practice. It was relatively inexpensive and very popular with farmers and legislators. Still, he once suggested privately that the state should charge for both *Hippodamia* and vedalia releases. "That would certainly have the effect of weeding out a lot of unnecessary applications," he wrote, "and incidentally would increase our working fund."[21]

Smith lacked the time or staff to do a thorough investigation of the *Hippodamia* program. In 1917, the U.S. Bureau of Entomology placed an agent in the State Insectary—something Howard was unwilling to do before Smith took charge—to cooperate on projects of interest to both agencies. The federal man spent three seasons on *Hippodamia*. He marked beetles with paint in the lab before release, then found that they flew away quickly rather than staying to help the grower who brought them in. Even the millions of beetles released by the Insectary were nothing compared to the existing population. They did not provide the hoped-for head start on the season's aphids. With this evidence, Smith at last could drop the program. He could not stop private operators from selling the beetles, which some farmers and gardeners have continued to buy over the years despite the repeated conclusions of entomologists that this practice does no good.[22]

As he cleared away some of the useless projects he had inherited, Smith resumed importation of potentially beneficial insects from abroad. In a trip nearly canceled because Governor Johnson opposed the expenditure, Smith sailed to the Orient in the fall of 1913, stopping in Hawaii to confer with the biological control workers there. He arranged for entomologists in Japan and the Philippines to collect and send beneficial insects, especially for citrus, and he brought back some material himself. With these contacts and others he was able to make because of his years in the federal service, Smith hoped to speed perishable parcels of insects through shipping channels, customs officials, and quarantine inspectors.[23]

The citrus industry naturally received top priority, having helped put Cook, and thus Smith, in office. Smith chose the so-called citrus mealybug (*Planococcus citri*)—actually a widespread pest of numerous field, orchard, and greenhouse crops—as the first new foreign project. He sent his first full-time explorer, Henry L. Viereck, to the Mediterranean in search of mealybug enemies in 1914. Viereck found one that came to be called the "Sicilian mealybug parasite," *Leptomastidea abnormis*. It received a lot of publicity in California and raised hopes of controlling *P. citri*. Smith warned growers not to expect the new parasite to control the mealybug completely. He nevertheless hoped that the new natural enemy might adapt to its new environment in a few years and prove valuable. *L. abnormis* quickly became common in the orchards, but did not solve the mealybug problem. World War I abruptly terminated exploration after only a few months, and Viereck left the California service. For the next two years, Smith had to rely on his network of correspondents, who produced little.[24]

In the meantime, Smith and his staff attacked mealybugs in a new way, reminiscent of the *Hippodamia* program but more challenging and also more effective. They tried mass production of a ladybeetle that preyed on mealybugs but failed to control them on its own. The cumbersome new approach had to be repeated annually, and it required the active participation of growers, but this may be why it solidified their support for biological control in the absence of a miracle like the vedalia.

An assistant in the State Insectary found that he could rear mealybugs on potato sprouts. This seemed a good way to support mealybugs and their enemies on long voyages from foreign countries, but it proved especially useful for breeding large numbers of parasites or predators prior to release in the field. On the appeal of citrus growers, Smith established a southern branch insectary at Alhambra, in the midst of the citrus belt and much more convenient than Sacramento. The Alhambra staff used potato sprouts, first to get the new Sicilian parasite going and then for mass rearing of the ladybeetle *Cryptolaemus montrouzieri*. This predator, introduced in the 1890s, often arose to wipe out large infestations of mealybugs. Smith's predecessors had boasted that the ladybeetle was as effective against mealybugs as the vedalia was against cottony-cushion scale. However, the "mealybug destroyer" recovered too slowly from winter to prevent pest outbreaks. Smith hoped that mass production would give *Cryptolaemus* a head start on mealybugs early in the season.[25]

The mealybug problem was getting worse at this time. Another, more destructive species, *Pseudococcus fragilis,* had appeared in 1913 and was spreading. Like other mealybugs and soft scale insects, the new pest produced honeydew that nourished a black fungus on the fruit. Growers faced their biggest crisis since cottony-cushion scale. In the space of three years, entomologists in South Africa, England, and California independently described the mealybug as a new species. The California description as *Pseudococcus citrophilus* came last and was therefore officially invalid, but the common name "citrophilus mealybug" stuck.[26]

Smith tabbed former San Diego County Horticultural Commissioner H. Morton Armitage to direct the southern operation, after a series of assistants had entered military service. Armitage refined the *Cryptolaemus* mass production technique. Temperature and other conditions, even the choice of potato variety, had to be carefully controlled in rooms full of spuds and insects, to insure maximum production of ladybeetles without overcrowding that would lead to cannibalism. Insectary workers collected the phototropic beetles from cloth hung in the windows. As the work took on industrial proportions, cost efficiency required an understanding of the best weather and time of year for release in each part of the citrus belt.

The citrus industry leapt vigorously into the *Cryptolaemus* program. The Limoneira Company of Santa Paula, managed by industry kingpin C. C. Teague, built its own insectary in 1916 to assist in *Leptomastidea* and *Cryptolaemus* production. Several local associations followed suit. Limoneira and other large outfits supplied materials to the state. Every county in the citrus region, beginning with Ventura in 1918, established a county insectary, partly to help the state rear and distribute newly imported beneficial species, but mainly to mass-produce the mealybug destroyers commonly referred to as "crypts." The most ambitious insectary of all, in Orange County, eventually grew to twenty-eight buildings. Some of this work came at the taxpayers' expense. Growers also contributed with assessments— voluntary or not, depending on the county—on fruit shipped. Many whose orchards remained free from citrophilus mealybug willingly helped in hopes of retarding its spread. To that end, state and county officials released beetles even in groves whose owners had not paid.[27]

"Crypt" factories grew through the 1920s as the pest spread across the southern California citrus belt. Upon taking over the state biological control program in 1923, the University of California left *Cryptolaemus* releases to county and private insectaries. By then, the beetles represented more of a commercial service than a research project. All involved considered it a great success. Smith and Armitage reported in 1931 that *Cryptolaemus* cost $5.60 per acre and was getting cheaper. Chemical treatment for the mealybug would cost $20–40 if it worked, which it did not. Armitage, who transferred to the university along with Smith and the rest of his staff, moved back to government service in 1924 to take charge of the Los Angeles County insectary program. Later he returned to the state Department of Agriculture. In 1944, Armitage became its chief entomologist.[28]

The mealybug program relied on crucial support from the organized citrus industry. Grape growers in the San Joaquin Valley also suffered from mealybugs, but the state would not run to their rescue unless they were prepared to help themselves. Smith suggested that grape growers provide the facilities and funds to get started and enlist their agricultural commissioners to bring in county assistance. Although one of the affected counties did build a *Cryptolaemus* insectary, valley conditions failed to sustain the ladybeetles. The state's fragmented agricultural bureaucracy included a board of viticulture separate from the horticultural commission. Because Smith worked for the latter, he had no direct jurisdiction over grape insects until the various agencies were combined in 1919.[29] Some cooperative arrangement could probably have been worked out had the grape industry or the viticultural board come up with the funds. Instead, Smith devoted most of his attention to other crops. Citrus growers remained his most eager customers, so he continued to put their needs first.

When foreign exploration resumed, therefore, citrus pests were again the target. Because the war made planned exploration in Europe and Africa impractical, Smith sent Curtis P. Clausen to China, Japan, and the Philippines in 1916. Clausen had worked briefly at the Citrus Experiment Station of the University of California, mostly on mealybugs, before joining the State Insectary. He began a long career of Asian exploration, although this first trip produced no stunning new natural enemies. Soon afterward, he joined the army. California could not match the salary that the U.S. Bureau of Entomology offered him for exploration after the war.[30] Clausen spent thirty-one years in the bureau and helped strengthen its ties with California, to which he would return in 1951 as Smith's successor.

Smith dispatched another explorer in 1917, not for citrus but for a northern California crop, sugar beets. Sugar interests, particularly the Spreckels Sugar Company, pressed Smith to try biological control against the beet leafhopper (*Circulifer tenellus*). Planter-backed efforts to control pests of sugar cane in Hawaii inspired this movement in California, where sugar production was similarly oligopolistic. The beet leafhopper was then thought to be native to North America, but native parasites failed to control the pest. Hoping that parasites of other leafhoppers might take to *C. tenellus,* Spreckels put up the money to send State Insectary assistant Everett J. Vosler to Australia. Direct sponsorship by corporate agriculture once more served biological control. The benefits, however, went again to citrus growers, not to sugar planters.

In two trips to Australia, Vosler found no parasites that could survive on beet leafhopper. While he was there, he pursued the ongoing mission of biological control of citrus insects. From George Compere's old haunts he sent *Metaphycus lounsburyi,* a parasite of black scale (*Saissetia oleae*). Both scale and parasite originated in Africa. California had been after *M. lounsburyi* since 1900. It may have been the species Compere tried to introduce at that time, when only the hyperparasite *Quaylea whittieri* reached California alive. Vosler's shipment established the desired parasite in coastal citrus orchards and inspired predictions that black scale was finished as a pest in California. Vosler did not live to see those forecasts fail. He entered the army, and died soon afterward of pneumonia following influenza at a southern California post.[31]

Smith believed that the long-sought *M. lounsburyi* could be the key to black scale control and free growers from $2 million in annual fumigation costs. He accepted the "sequence theory" of his former USDA supervisor William Fiske. Fiske had argued that control of an insect might require a parasite of each major developmental stage. The new black scale parasite filled a gap in the sequence, between an egg parasite and a predator that had been introduced many years before with great fanfare but disappointing results.

Having committed most of his resources to mass production of mealybug predators, Smith went to C. C. Teague for help with *Metaphycus lounsburyi*. Already the most powerful member of the California Fruit Growers Exchange, Teague was soon to begin his thirty-year presidency of the cooperative. The Exchange granted

$3,000 to be matched by state funds. Teague held out some fifty trees at Limoneira from fumigation as a test site for the new parasite in 1919. After the initial results showed promise, Teague expanded the test plot to ten thousand trees the next year. Smith assigned assistant Harold Compere, son of the former explorer George Compere, to supervise the operation and a similar test at Alhambra, an inland site.[32]

The parasite failed to control black scale in the Alhambra experiments. Under inland citrus conditions, black scale produced only a single annual generation, popularly called "even-hatch." Most of the parasite population starved during half of the year when there were almost no scale of the appropriate stage. In the coastal conditions of Ventura County, overlapping generations of black scale could more easily sustain the parasites. Harold Compere could not temper the enthusiasm of Limoneira and Exchange officials, who believed that *M. lounsburyi* would soon free the entire citrus belt from black scale.

The Exchange hired its own entomologist, Russell S. Woglum, in 1921 to demonstrate fumigation techniques, his specialty. As a USDA citrus entomologist, Woglum had previously come into conflict with the State Insectary, criticizing Smith's corps as too technical and impractical. Now Woglum began to spread *M. lounsburyi* through the interior regions for a penny apiece. Smith accused Woglum of jumping ahead of research on the value of mass production for the interior in order to boost his own popularity with growers. This, Smith complained, would discredit biological control. Smith and his assistants feared losing their authority over biological control to the county commissioners on the one hand, or to the Exchange on the other. Limoneira's assistant supervisor wanted Harold Compere fired for writing an article that rejected *M. lounsburyi* for even-hatch black scale.[33] Industrial patronage had its drawbacks.

This wave of enthusiasm passed after the third season. Woglum's work in the even-hatch areas failed completely. The hyperparasite *Quaylea whittieri* became numerous in coastal areas and was at first blamed for inadequate control there. *M. lounsburyi* may have facilitated fumigation by evening up the hatch in some coastal areas, but in doing so the parasite prevented its own complete success. Mass production of the black scale parasite failed to duplicate the performance of *Cryptolaemus* against mealybugs. Black scale remained a major target for Smith's explorers.[34]

Smith asked L. O. Howard for advice on hiring a new explorer in 1919. Lest California return to careless enthusiasm for biological control, Howard emphatically rejected the prospect of George Compere going out again and recommended that Smith accept the application of Everett W. Rust. A former assistant at the Whittier lab of the University of California, Rust had spent several years working on cotton insect parasites in Peru. Howard characterized him as too restless for a fixed position, but a good choice for foreign exploration.[35]

After a short stay in Australia, Rust moved to South Africa, where he would remain until 1927 except for one trip home. For two decades, South African entomologist Charles P. Lounsbury had pleaded with state and university entomologists to come and find the enemies that, he was sure, kept black scale under control in that country. He and his own staff could not do the searching themselves. They had no time to worry about an insect that caused their growers no trouble. Lounsbury, a transplanted American, agreed that the excesses of the old State Commission of Horticulture had not served biological control well. But he blasted critics who could have channeled the board's enthusiasm to good effect through cooperation. Lounsbury singled out Berkeley entomology professor Charles W. Woodworth. Woodworth never did participate in biological control, but others in the university began to study parasitic insects and to associate with the program of the State Insectary.[36]

Rust found an assortment of black scale parasites, some of which the staff back in California described as new species. Little of what he shipped, however, arrived alive. Only two out of dozens of natural enemies were established, without much practical effect. Some species whose peculiar habits were yet unknown failed to survive in the lab. Others that Rust considered unimportant in controlling black scale in South Africa were never sent at all. Smith criticized Rust's low productivity as early as 1921, but left him on the job. Rust apparently did not enjoy his work. Some of his many complaints were understandable. Shortly before Smith recalled him in 1927, Rust suffered

attacks of dengue fever and wrote, "About the second day I was afraid
I was going to die, and the next day I was afraid that I wasn't." He
survived, but he was glad to leave. True to Howard's assessment, Rust
could not endure confinement to a lab. Instead, he took a job in
Hawaii in 1928, following a long line of California workers who had
moved to the islands.[37]

Meanwhile, Smith's biological control program solidified its
commitment to citrus and moved from government
bureaucracy to the University of California. Early in the
century, university entomologists had severely criticized the horticultural commission and its narrow-minded reliance on natural enemies.
Charles Woodworth, the first full-time entomologist at Berkeley,
perceived far greater ecological complexity than the zealots would
admit. That awareness also led Woodworth and his growing staff to
complain that chemical salesmen had too much influence on farmers.
Woodworth called for spraying only at times and places proven
effective. He warned that too much treatment could do more harm
than good. By the time Harry Smith established a working relationship
with both Berkeley and the Citrus Experiment Station at Riverside, the
university was not completely averse to biological control. Riverside
entomologist Henry J. Quayle took an interest in parasites, although
he shared some of Woodworth's criticisms.[38]

Smith yearned for more than a working relationship with the
university. State government work frustrated him as a scientist. He
had wanted an academic position since his early days in the USDA.
He believed that government entomologists should concentrate on
economic ends, but those ends required scientific research. Smith grew
restless after only a couple of years in California. He rarely had time
for close biological study. He complained when governors refused to
send state scientists to scientific meetings. Smith hoped to return to the
USDA if his prospects in California did not improve. Administrative
duties increasingly encroached on his time, especially after the

California Department of Agriculture was formed in 1919 to unite several existing agencies. The director of the department put Smith in charge of all pest control activities, including regulation of chemicals and even work on snails and rodents. Smith wanted to move all biological control activities to southern California and work on them full time, but the director denied his request.[39]

Smith thought a transfer to the University of California would benefit biological control and his own career, freeing him from the bureaucratic work he had to do in Sacramento. It is not clear whether he suggested the move, as was reported in later years, but he certainly approved of it. When the time came in the spring of 1923, Smith drafted the bill that transferred work with beneficial insects from the department of agriculture to the university. The driving force was not Smith, but C. C. Teague. The powerful president of Sunkist exerted considerable influence over the university and, through the Agricultural Legislative Committee, over state government. He wanted Smith to work on biological control full time. Smith hoped to move to Berkeley and be part of a larger scientific community than was available in southern California. Teague, however, invited university administrators to his ranch and persuaded them to attach Smith to the Citrus Experiment Station.[40]

Smith and four assistants joined the staff of the University of California. At Riverside they comprised a Division of Beneficial Insect Investigations, distinct from the Division of Entomology. He had to fight for his title of associate professor. Smith, then thirty-nine, believed that his experience in the USDA and state government merited a full professorship, but administrators refused at first even to grant him tenure. A few years later, they tried to take away any professorial title and leave him as merely "assistant entomologist in the experiment station." A vice president argued that someone who did not teach regular courses—meaning almost anyone at Riverside—was not really faculty. Smith countered that his work in summer courses and with graduate students justified his title.[41]

The transfer preserved the program's institutional identity. Had it been merged into entomology, biological control might have dwindled over the years, as it did in the federal government. Changing research priorities within a single administrative division might have relegated the method to a secondary role in which it could not survive. Teague's insistence that Smith's team come to Riverside also helped to shape the future of biological control. It would continue to have a willing constituency in the citrus growers, who had supported it since the vedalia beetle swept cottony-cushion scale aside, and who now eagerly sponsored mass production of mealybug predators.

Nominally a statewide service, what now became the Division of Beneficial Insect Investigations had always focused on citrus. In 1923, the team joined like-minded scientists from several disciplines, more than four hundred miles from the diverse Berkeley campus. Catering to the citrus industry would retard the expansion of biological control into other crops, but would keep the program alive until such expansion began in earnest in the 1940s.

Smith's faculty status provided a key to that expansion, because he could offer the only graduate training in biological control in the United States. Students and colleagues alike affectionately called him "Prof Harry." Upon completing their training, several of his students joined him on the staff. After ten years with limited time for research, Smith began to flex his muscles as a scientist, although he always believed that his first duty was to agriculture.

That belief was tested soon after Smith moved from the shadow of the state capitol in Sacramento to the arid hills overlooking Riverside. One of his first tasks was to hire a replacement for Morton Armitage, who preferred direct service to growers (as a county insectary director) over university research. Smith wanted an insect parasitologist who could discover why many potentially valuable species failed to survive in the lab after shipment from abroad. Philip H. Timberlake, an old friend and colleague from gypsy moth days in the U.S. Department of Agriculture, seemed to be the right choice because of his experience with hymenopterous parasites of scale insects in California and Hawaii. Smith managed to match the salary the Hawaiian Sugar Planters' Association was paying Timberlake in 1924. The Maine native became one of the few biological control figures to move *from* Hawaii *to* California. The fact that his wife (the aunt of future president Richard Nixon) was from southern California helped make the move attractive.[42]

Timberlake gained notice in later years for staying at his microscope and working as hard as ever, on meager retirement pay, well into his nineties. Beginning soon after his arrival, he focused that dedication on the taxonomy of wild bees, not on biological control. He stubbornly refused to work on pests and parasites. He did, however, retain a legendary skill at identifying parasites brought to him—always grunting about being disturbed from his microscope. On a trip to Washington late in his life, colleagues recalled, Timberlake was more interested in the Smithsonian Institution collection of bees than in his nephew's inauguration. Although the tenured Timberlake kept his job, Smith denied him promotion for the rest of his career. In Smith's view, the experiment station existed for the benefit of farmers, not scientists. Research not turned toward practical ends was an inappropriate use of tax dollars earmarked for agricultural science. This attitude, of course, endeared Smith to the citrus industry.[43]

Having only one faculty member, the division functioned more as a team with a captain than as a collection of individual scientists, even if Timberlake refused to play his assigned position. The team approach gave California a great advantage over the USDA in biological control work. The Bureau of Entomology launched major new projects against the European corn borer (*Ostrinia nubilalis*) and Japanese beetle (*Popillia japonica*) after World War I. Bureau chief L. O. Howard discussed the possibility of creating a new division for biological control, but never did it. One division handled the corn borer, another the Japanese beetle. Division leaders, who jealously guarded personnel and funding, were not disposed to cooperate with each other or to assign natural enemies a high priority.

Apathy toward biological control, on the part of top brass as well as the field agents responsible for evaluating results and talking to farmers, dismayed the corps of explorers stationed in Europe and Asia. By the 1930s, the two big projects degenerated into perfunctory shipment of the same few parasites year after year, without achieving control of either pest.[44]

The one bright spot in USDA work during the period was the California-trained Curtis Clausen's conquest of citrus blackfly (*Aleurocanthus woglumi*) in Cuba in 1929-31. The Cuban government paid the expenses for biological control by the Bureau of Entomology, which wanted to keep the insect from reaching Florida. Clausen and his assistants went to Malaya and soon found a parasite that completely suppressed the blackfly in Cuba. Years after Howard's retirement, a new bureau chief made Clausen head of a Division of Foreign Parasite Introduction, to coordinate exploration for natural enemies of any pests.[45] Although exploration became more efficient, the other divisions still took over once new parasites passed quarantine, despite Clausen's recommendation that his division be given responsibility for all phases of biological control. The lack of teamwork or contact with farmers continued to hinder federal efforts.[46]

Smith had a smaller staff than the USDA, but his team covered all aspects of a biological control project. His success rate in the 1920s and 1930s was no better than that of the federal entomologists, yet his program grew while that of the USDA crumbled. Higher morale and frequent contact with appreciative growers kept biological control afloat in California. Smith reduced the intensity of propaganda, while continuing to remind growers of the great success against cottony-cushion scale. Unlike his USDA counterparts, who pursued the same parasites for decades at a time, Smith was able to move on to new projects quickly, especially after joining the University of California.

The university left mass production of the mealybug predator *Cryptolaemus montrouzieri* to county and private insectaries. With that situation seemingly in hand, and with continued hope for black scale parasites from Africa, Smith turned his attention to armored scale pests of citrus, in particular to California red scale (*Aonidiella aurantii*) and purple scale (*Lepidosaphes beckii*). These scalf inàcts moved to the top of the citrus growers' list, and thus to the top of Smith's. Red scale, especially, would remain a high priority for the rest of his career.

Smith hired another old friend for exploration. While in Italy in 1911 on USDA work, Smith had spent time in the laboratory of biological control enthusiast Filippo Silvestri. Silvestri made himself available as a free-lance explorer, and once worked on Mediterranean fruit fly for Hawaii. In 1924, Smith hired the Italian entomologist for a year and half to search for red and purple scale parasites in the Orient. Hampered by the civil war in China, Silvestri sent only one parasitic species that Smith's team was able to establish in California, and none on California red or purple scale. California did not use Silvestri's services again. Entomologists years later concluded that he, like George Compere before him, sent useful parasites that were discarded upon receipt. At the time, they were thought to be identical to species already in the state.[47] Sorting such species out became a major part of the division's work in the 1930s. (See Chapter 5.)

Before Smith's staff could resume efforts against armored scales, the citrophilus mealybug returned to prominence in areas where *Cryptolaemus* releases had satisfied growers in previous years. Some growers and county insectary workers said that weather had turned to favor the mealybug. Although fruit could be harvested undamaged, it all had to be taken relatively early in the season. The resulting gluts and low prices disrupted a market that the industry had carefully regulated by virtually year-round orange production. Growers demanded a solution from the Citrus Experiment Station. Many installed high-pressure water spray rigs to blast mealybugs from the trees if nothing else would work.[48]

Earlier explorers had not found citrophilus mealybug in Asia, the native home of most citrus species. Smith chose Australia as the next most likely source of the pest. Tabling black scale work for the moment, he called the unhappy Rust home from Africa and sent Harold Compere to Australia in 1927. The young Compere, who had grown up listening to his father's travel tales, was fairly adventurous in his own right. Stationed in Texas as an army pilot at the end of World War I, he had been disappointed not to get to Europe. At that time, the USDA was trying to establish a cotton-free belt to retard the spread of pink bollworm. Compere thought of using air surveillance to enforce the no-cotton zone, and proudly took part in the first agricultural use of an airplane.[49]

In Australia, Compere traced the steps Albert Koebele had taken thirty-nine years before, and laid the basis for a career in entomological exploration. Compere soon found citrophilus mealybug with parasite exit holes, indicating that he had come to the right place. Getting live parasites back to California, however, was more involved. Botanic gardens proved better hunting grounds than orchards, where the mealybug was hardly to be seen. A suburban mulberry tree, heavily infested with mealybugs one month, became a prime source of parasites the next.

As Compere reasoned, an upstart pest population in such a spot could build up before natural enemies found it. In turn, the natural enemies would attain large numbers before cleaning up the outbreak. Then both would again become scarce. Cottony-cushion scale and its enemies often followed this pattern. Indeed, Koebele had taken the greatest numbers of vedalia beetles in areas they had just reached for the first time. A parasite hunter coming upon the right spot at the right moment had an ideal opportunity to collect a species that otherwise, due to its own effectiveness in controlling its food source, might be too rare to find. "Sidewalk exploring," rather than tramping through the wilds or even the orchards, became Harold Compere's principal method on subsequent expeditions.

Compere reared six species of mealybug enemies in his makeshift Australian laboratory. One reached the Riverside quarantine room on branches cut from the mulberry tree with the owner's permission. Others required careful nurturing of potato sprouts to maintain mealybugs and parasites all the way home by ship. Even that procedure would have been impossible had Compere not slipped contraband American potatoes into his lab with the cooperation of an Australian quarantine officer. Two species of parasites and three of predators made it to California alive.[50]

The young explorer scored one entomological triumph even before the parasites were released into California orchards. In 1927, Smith was embroiled in a dispute with Stanford University entomologist Gordon F. Ferris, the state's leading scale insect taxonomist, whom the Riverside group frequently consulted for identifications. Ferris respected the work of precious few entomologists. In print, he criticized some of the foreign work in Smith's division as ill-conceived. While Compere was searching in Australia for parasites of

Pseudococcus fragilis, Ferris dismissed the mealybug on that continent as another species, on the basis of samples Compere sent.

Compere, a high school dropout, dared to insist that he could correctly distinguish the two mealybug species while the outspoken Stanford professor and topnotch coccidologist could not. That Compere prevailed was most satisfying to Smith and the Riverside group.[51] The ugly rivalry between early California officials and the USDA had served no good purpose, but this more friendly competition with Ferris produced valuable results and bolstered the confidence of the biological control team.

A greater victory occurred in the citrus groves. County insectaries helped to rear and release the new parasites, *Coccophagus gurneyi* and *Hungariella pretiosa,* which multiplied so fast and became established so easily that the three predators could be ignored. In two years, the mealybug nearly disappeared from California. Harold Compere found greater success in his first time out than his father had ever known.

Insectary operators, whose *Cryptolaemus* mass production business was threatened, denied the success of Compere's parasites. Morton Armitage of Los Angeles County claimed that a climatic cycle had driven the mealybug down temporarily. Smith was diplomatic, but Compere enjoyed watching the unbelievers squirm. As the years passed and the pest did not regain its former status, all had to admit the greatest success of biological control since cottony-cushion scale. Indeed, although publicity never matched that for Koebele's achievement, control of citrophilus actually surpassed it. Cottony-cushion scale still broke out occasionally following pesticide use, but the mealybug did not. The new parasites worked on their own, with no need of annual release. The role of county insectaries diminished to assisting the establishment of newly introduced species and producing some *Cryptolaemus* for other mealybugs less threatening than citrophilus.[52]

Neither parasite had much territory to itself before the other moved in, so there was no way to tell whether one would have been sufficient. Smith and Compere concluded that the two species complemented each other and deserved credit for the complete subjugation of citrophilus mealybug. The effect of a combination of parasites reinforced theoretical ideas Smith was forming on the general question of multiple natural enemy introductions. This was his first serious move into theoretical aspects of biological control, although he had contributed to its terminology while still in state government. Smith no doubt wanted to be recognized as a scientist as well as a public servant. Agricultural productivity remained his primary goal, but even at the Citrus Experiment Station the university provided more of an academic atmosphere than he had known in Sacramento. He wanted to be a part of academic science. It gave him another audience, one quite different from citrus growers. Still, only direct challenges to the practical usefulness of his work spurred him to publish his theoretical ideas.

S mith came to population dynamics theory with the common assumption that populations maintained relatively constant, equilibrium levels under normal conditions. According to the most basic premise of biological control, a pest was a pest only because conditions became abnormal. About half of the agricultural pests in the United States were foreign species that had presumably escaped from their natural enemies. Unlike the partisan enthusiasts who preceded him in California, Smith acknowledged the other half, species that became too abundant even in their native land. But these, he argued, were also abnormal. Agricultural monoculture changed their environment by providing an overabundance of food.

The adherence of Smith and his staff and students to the idea of "balance of nature" has been derided as dogmatic. Perhaps balance really was "an unquestioned article of faith." Its widespread acceptance helped fuel the rise of economic entomology as a profession. However, Smith's commitment to natural balance must not be mistaken for a similarly unyielding belief that classical biological control could solve all pest problems. The latter view belonged to Smith's California predecessors, but not to him or most of his successors. Smith and his colleagues adjusted their theoretical details when data required it.[53] Nevertheless, the notion of populations in equilibrium remained so central to the Californians, even as it came under greater attack from the 1930s onward, that they focused their attention only on examples that supported their general views.

Smith did not advocate changing cultural practices as a pest control measure. He was as committed to large-scale farming as he was to natural balance. For an economic entomologist, to suggest that farmers grow less of their most profitable crop was to surrender. But all pests constituted a tiny fraction of insect species. Most species remained rare in their normal environment. Therefore, Smith reasoned, their populations were stable.[54] Smith and others adopted the term "density," which included not only the raw size of a population but also its distribution.

The stability of densities was a fundamental assumption. Nobody was out counting all of those rare species to see if their numbers really remained constant. If the population density of some insect fluctuated by several orders of magnitude, but never reached pest status, economic entomologists like Smith could only call it "rare." Although Smith knew that real numbers did fluctuate, he considered those fluctuations narrow enough that the *average* density was a true equilibrium value with biological significance.

The first elaborate mathematical approach to the relationship between an insect and its natural enemies violated this assumption. William Robin Thompson, a former colleague of Smith's in the gypsy moth program, took charge of USDA exploration for natural enemies of the European corn borer in France in 1919. Thompson constructed a set of equations to describe changes in the numbers of hosts and parasites. During the war, he had become enthusiastic for mathematical description of biological phenomena. He had previously exasperated colleagues by gathering quantitative data instead of live insects.[55] In France, Thompson worked closely with Paul Marchal, who had devised a crude model for sudden population crashes brought about by parasitism. Thompson treated reproduction as a constant rate and assumed, as Marchal had, that each parasite found a separate host, so that superparasitism did not occur.

Thompson designed his equations to calculate the number of generations between the introduction of a parasite and the crash of the host to virtual extinction. He calculated that such a crash would occur if the parasite's reproductive rate exceeded that of the host. During the intervening generations, the host would continue to increase, disguising the buildup of parasites until that generation when complete control suddenly occurred.

Thompson considered virtual eradication, rather than a low but persistent equilibrium pest population, to be the goal of natural enemy introduction. Ignoring the fact that even Marchal thought such dramatic cases exceptional, Thompson excitedly declared that he had found the theoretical basis for biological control. In practical terms, his discovery should mean that parasite introductions whose impact seemed negligible were in many cases not failures, but smashing successes still in the making. Success took so long only because parasites were introduced in such small numbers, compared to the pest population. Thompson's chief, L. O. Howard, did not understand the math, but nevertheless took comfort in this assurance that biological control, though slow, was on its way in many cases. Thompson tinkered with formulae throughout the 1920s, but he retained his general scheme for population crashes. His American superiors balked at Thompson's mathematical musings, so Marchal helped him get them published in France.[56]

Other economic entomologists were also wary of Thompson's mathematics. Although an increasing proportion of them were college-educated and even had advanced degrees, few possessed the background to plow through Thompson's equations. He admitted that his formulae neglected some of the complexity of real ecological systems, but he considered his conclusions sound. The factors not yet mathematized might affect the number of generations to extinction, but it should still occur.

Harry Smith stood among those less mathematically inclined who feared that theory had carried his friend Thompson away from biological reality. Those who most enthusiastically took up the mathematization of predator-prey relationships in the 1920s were not biologists, but mathematicians. Alfred J. Lotka and Vito Volterra drew analogies from physics to derive equations for continuing, periodic oscillation of predator and prey populations.[57]

Smith believed in steady equilibrium, not sudden crashes. Pondering what could maintain that equilibrium, he read the population dynamics articles of Thompson and others, although he could not understand the calculus that Lotka and Volterra used. Smith resisted their enthusiasm for mathematics not because it could never describe biology, but because it did not yet do so. Specifically, Smith objected to Thompson's assumption that the parasite population increased at a constant rate. As equilibrium approached, Smith wrote, rates of increase for both host and parasite must change and approach zero.[58]

During the 1920s, Smith developed his view of equilibrium as a balance of the opposed forces of reproduction and "natural control." The latter included all causes of death: starvation, climatic phenomena, and natural enemies. In Smith's experience, insects rarely reached the point of starvation before other elements came into play.[59] Of the remaining factors, only natural enemies increased with the abundance of the insect in question. This was just the conclusion of L. O. Howard and W. F. Fiske, whom Smith revered and with whom he had worked in the USDA.

One rare entomologist who did not resist the movement into mathematics was Royal N. Chapman of the University of Minnesota. Chapman's studies of flour beetles were among the first laboratory experiments on insect populations. Like Smith, he lacked a strong mathematical grounding, but he saw quantification as the ticket ecology and entomology needed to attain the scientific status of physics and chemistry. Chapman emphasized stability over population crashes, and made an analogy between electricity and population balance. A species' "biotic potential" encountered "environmental resistance" the way voltage related to electrical resistance. He defined the phrases rather vaguely, but biotic potential essentially meant the capacity for growth and reproduction under ideal conditions. Environmental resistance included both abiotic forces and natural enemies. Chapman apparently did not consider it to vary with population density. Rather, resistance in a particular environment was a constant against which reproduction would strike an equilibrium.[60]

Smith applauded Chapman's effort to stimulate the interest of entomologists in mathematics, and recognized the similarity between Chapman's concepts and his own. Smith quickly accepted Chapman's phrase "environmental resistance" and took it to mean all causes of death or reduced reproduction. Instead of Chapman's imprecise "biotic potential," Smith continued to use "reproductive capacity," the sheer number of offspring possible under the hypothetical circumstances of unlimited resources and no environmental resistance. Environmental resistance, then, must maintain populations at equilibrium density.[61] Smith still lacked a concrete way to express how it did so. The issue of introducing more than one species of natural enemy to control one pest sharpened his thinking in the late 1920s.

Tracing the results of Hawaiian introductions against the Mediterranean fruit fly (*Ceratitis capitata*), USDA agents Cyril E. Pemberton and Harold F. Willard had opened the debate in 1918. Pemberton and Willard declared that the use of more than one species of parasite caused poorer control of the fly than would have resulted from the best species alone. This conclusion stemmed from what happened when two species, *Opius humilis* and *Opius tryoni*, attacked the same medfly larva. In such cases of multiple parasitism, the developing *O. tryoni* larva usually killed its competitor. *O. humilis*, however, parasitized a higher percentage of the host population and therefore seemed the "best" parasite. Pemberton and Willard reasoned that the less valuable parasite benefited the fruit fly population by inhibiting the work of the better species.[62]

California biological control policy, like that of Hawaii, called for trying all possible natural enemies of a pest until it was under control. The conclusions of Pemberton and Willard directly challenged that policy. Smith thought on the matter for a decade before publishing a rebuttal, in his first theoretical paper, in 1929. He argued that the *individual* phenomenon of multiple parasitism would not automatically inhibit control, which was a *population* phenomenon. What mattered was which parasite could keep the pest population at the lowest level. This would be the parasite that best thrived in the greatest number of ecological circumstances, or niches, where the pest lived.

Multiple parasitism occurred rarely in the population, and hosts attacked by more than one parasite were just as dead as those attacked only once. If the species that lost such battles filled niches the other could not, Smith argued, the host population would be lower than if either parasite worked alone. Otherwise, interspecific competition displaced the loser from the field, with no benefit to the host. Overall, destruction of the pest would equal that by the best enemy, plus contributions by others if they were superior in any of the niches available in the region. Turning to the medfly case, Smith pointed out that while the introduction of *O. tryoni* reduced the percentage parasitized by *O. humilis,* total parasitism had increased. The Hawaiians' own data showed as much. In fact, before studying multiple parasitism and becoming alarmed, Pemberton and Willard themselves had suggested that the various fruit fly parasites would each predominate in different niches, presumably giving better overall pest control.[63]

Smith tied competitive exclusion to the question of equilibrium population densities. The "best" natural enemy in a given niche maintained the host at the lowest density, preventing an enemy requiring a higher host density from surviving at all. This part of the argument did not appear in his 1929 paper, but was in Smith's mind by 1927. He wrote privately that displacement of one natural enemy by another in a given situation could occur only if the newcomer were the more effective in control.[64]

Smith must have reached his conclusions at least in part from considering the most familiar case of biological control, which he had observed ever since arriving in California in 1913. At the time Smith was working out his argument for multiple introductions (1927–29), William H. Thorpe, an English postdoctoral fellow in his lab, studied the cottony-cushion scale and its introduced enemies. Thorpe found the often overlooked parasitic fly *Cryptochetum iceryae* to be just as important as the familiar vedalia beetle. The two divided the territory in California according to climatic differences. Control of citrophilus mealybug by two parasites only strengthened Smith's defense of multiple introductions. After 1929, other biological control leaders—including Curtis Clausen of the USDA, and Robin Thompson, who moved from the USDA to British Commonwealth service in 1928—accepted Smith's arguments. Work went on as before.[65]

The beneficial effect of intraspecific competition replaced W. F. Fiske's "sequence theory" as justification for multiple introduction. Successful biological control, where it occurred, generally involved only one or two natural enemies, rather than a sequence covering each developmental stage of the pest. Thompson explicitly rejected the sequence theory in 1923, and Smith stopped citing it at about that time. Although their views of how natural enemies achieved biological control contrasted sharply, both Smith and Thompson came to believe that introduction of many natural enemies was the best policy because it increased the chances of getting the one or two that could do the job.[66]

Smith's response to the multiple parasitism question rested on his assumption that a natural enemy maintained its prey at an equilibrium density. He turned more toward mathematics in an effort to explain how natural enemies did this. He might never have published on the subject, however, had others not challenged the equilibrium concept and the validity of biological control.

The threat came from entomologists Boris P. Uvarov, a Russian transplanted to Britain, and Friedrich Simon Bodenheimer, a German Zionist transplanted to Palestine. Uvarov made his reputation on studies of migratory locusts in Russia. In 1931, he published a book-length review of insect mortality due to climate, and explicitly denounced the idea of population equilibria. Uvarov argued that physical factors killed many more insects than biotic factors did, and must therefore be most important in determining population size. Average numbers, Uvarov declared, were nothing but a statistic, without biological significance. Populations fluctuated irregularly, communities changed, species evolved. "The conception of a normal number is a fiction."[67]

Bodenheimer also asserted the primacy of climatic factors, but spoke directly to the issue of applied biological control. He had tried to introduce the ladybeetle *Cryptolaemus montrouzieri* into Palestine against citrus mealybug. As in California, the predators built up too slowly each season to prevent crop damage. Bodenheimer concluded that introduction of natural enemies could not control pests in any environment with distinct seasons. He did not deny the existence of balance. Like Uvarov, however, he rejected what he perceived to be a nearly universal belief in its maintenance by biotic factors.[68] After publishing this view, Bodenheimer challenged Harry Smith's prized new triumph over citrophilus mealybug.

Bodenheimer stopped in Riverside on a trip around the world in 1931. He spent most of his time with Sunkist entomologist Russell Woglum, one of those who doubted that the new parasites from Australia controlled *Pseudococcus fragilis*. Bodenheimer challenged Smith to prove the effect statistically. Smith had an assistant collect data for several months, but nothing came of it. He had to rely on the permanent subsidence of the mealybug as evidence of the parasites' work, as he also relied on repeated instances of cottony-cushion scale reduction all over the world after vedalia beetles were introduced. This was proof enough of biological control. In succeeding years, Smith and his students looked for other ways to demonstrate the effect of natural enemies on pest populations. Smith himself concentrated on population theory. His perspective was always that of a biological control administrator, with the case of cottony-cushion scale ever in mind.[69]

Bodenheimer's experience in harsh, seasonal environments no doubt influenced his opinion on the role of climate. Sudden swarms of locusts, such as Uvarov studied, appear as far from equilibrium as any system imaginable. Smith, meanwhile, spent his whole career trying to adjust equilibria below economic damage levels. Population balance was his job. In the system he knew best, that of cottony-cushion scale, the permanent effect of natural enemies was beyond question. Although he did not participate in the original project, he often had occasion to observe this system in action and to refer to it in speech and print. No citrus insects in California swarmed and crashed with the conspicuous violence of locusts. It is neither new nor surprising to find that the particular insects and environments an entomologist knew best could largely determine his or her generalizations.[70]

Bodenheimer challenged more than Smith's experience. Professional outlook and pride were also at stake. The control of citrophilus mealybug by newly introduced parasites cemented Smith's faith in population balance and in the ability of natural enemies to determine equilibrium levels. He feared that Bodenheimer and Uvarov had damaged economic entomology, presumably by disparaging biological control.[71]

Responding directly to Bodenheimer, Smith at last published his own reasoning on population dynamics in 1935, although most of the ideas were in place several years earlier. Students recalled Smith's mathematics as a hobby that he pursued intensively, but at which he did not excel. His argument was largely a qualitative interpretation of the work of Vito Volterra and Raymond Pearl, with respect to biological control. Smith examined the logistic equation, an expression for population growth and equilibrium. The equation, which Pearl rediscovered and attempted to verify in the laboratory, predicted that a growing population would increase geometrically and then level off. Pearl's single-species systems roughly matched this pattern. Although no natural enemies were involved, other effects retarding population growth clearly intensified with population density.[72]

Smith noted that studies of logistic growth followed the path from low density *to* equilibrium, whereas economic entomologists also wished to maintain equilibrium *after* it had been reached. The factor Smith wanted to analyze was predation (or parasitism). He already knew that, to be decisive, it must intensify with population increase and relax with decrease. Smith emphasized this simple, logical deduction about equilibrium as his most important point. His mentors Howard and Fiske had reached the same conclusion, but had not specified the mathematical relation between mortality and density. Volterra, treating predatory attacks as random encounters analogous to those between gas molecules, assumed that the proportion attacked would increase linearly with density. Volterra's discussion was good enough for Smith, who graphed proportional mortality as a linear function $m = aD + b$, where D stood for density (see Figure 4.1).

Mortality was thus the sum of primarily biotic factors, such as natural enemies and intraspecific competition, and another set not related to density. The latter consisted mainly of climatic factors.

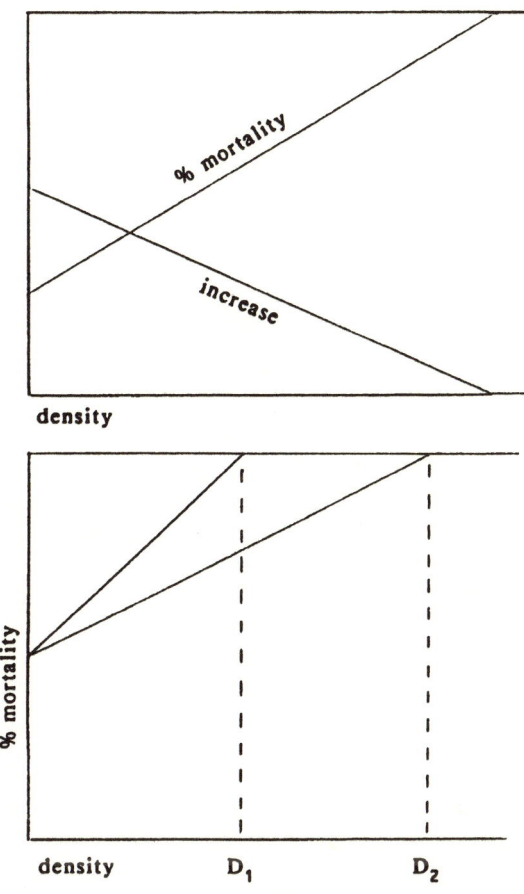

FIGURE 4.1. Harry S. Smith's schematic representation of density-dependent mortality. In a hypothetical population undergoing logistic growth, the mortality rate is a linear function of density: $m = aD + b$. Population increase drops to zero at equilibrium. The lower graph shows the effect of different environmental conditions on equilibrium density and on the magnitude (slope) of the density effect. D_1 and D_2 are equilibrium values under different sets of conditions.

Smith proposed to name the former "density-dependent" and the latter "density-independent." The labels corresponded to what Howard and Fiske called "facultative" and "catastrophic," respectively. Even W. R. Thompson, who did not include a density reaction in his mathematical formulae, identified similar categories. Smith's terms prevailed, although he was not a prominent participant in later debates over these concepts.[73]

According to Smith, Bodenheimer's claim that climatic factors could "regulate" population densities failed by logical reasoning. Those factors could not increase their proportional kill in response to increasing density, or relax in response to decreasing density. Therefore, their long-term effect in the absence of other factors must be either extinction or infinite increase. "Regulate," Bodenheimer's word, became the catchword for Smith and his followers, and only density-dependent factors could regulate. Bodenheimer denied the obvious.[74]

Bodenheimer's mistake, in the eyes of Smith, was to equate the proportion of deaths from any one factor with its importance in regulation. Experience told Smith that in practical biological control a high proportion of attack meant nothing by itself. All too often, high parasitism rates raised hopes for control that never occurred. By the late 1920s, Smith gave up on percentage of parasitism as a measure of a natural enemy's economic value. If the hosts killed would have died from other causes in the absence of parasites, then the parasites were not decisive. The same must apply to climatic mortality. The raw share of kills did not matter; what did was the factor's ability to increase that share when the pest population grew. Bodenheimer's data on climatic factors did not address this, whereas entomologists frequently observed a higher percentage of parasitism at higher host densities. Smith found Thompson's criterion of high reproductive potential to be no better way to judge parasites. It might determine how quickly the system reached equilibrium, but not where that equilibrium value lay.[75]

Consideration of equilibrium, and the failure of the most obvious criteria for evaluating a natural enemy, led Smith to ask what the right criteria really were. How did a natural enemy maintain its host at a steady density? Why did one parasite or predator do a better job than another? Persistence of the natural enemy species, and increase in response to host density, depended on *actual* reproduction, not potential. Actual reproduction depended—indirectly for predators, but directly in the case of insect parasites—on finding prey. The species that did this best would maintain the lowest prey density. The process by which the adult female parasite found a host for oviposition had fascinated Smith since very early in his career.

In the multiple parasitism paper of 1929, Smith wrote that the best parasite species was the one that could maintain its own population at the highest equilibrium level by finding the most hosts. This conclusion drew on a crucial part of the density-dependence concept: Host and parasite regulated each other's populations. By 1933, he turned the argument around. The best parasite kept its own numbers *lowest* by its ability to find hosts that were far apart—that is, at low density. Through interspecific competition, a "good" natural enemy displaced a poorer one by leaving hosts too far apart for the poorer searcher to survive between encounters. Since the "best" would always win out, entomologists need not fear harm from introducing more than one parasite. The principle of searching ability also explained the density response. At a higher prey density, a greater number of prey fell within reach of each individual enemy, increasing its nourishment and thereby its actual reproduction. Increased reproduction meant more natural enemies, which in turn increased the proportion of prey killed.[76]

After he had made these arguments, the appearance of a textbook ignoring them all naturally disappointed Smith. Harvey L. Sweetman published *The Biological Control of Insects* in 1936 to use in his own course at the University of Massachusetts. Sweetman never attempted biological control himself, but he wanted to teach all aspects of economic entomology. Finding no text, he compiled one from the literature. Smith planned to write a biological control book, but never got it done. Sweetman's remained the only text until the 1960s. Although he corresponded with Smith and cited him on restraining exaggeration of biological control, Sweetman's limited discussion of theory depended entirely on Thompson. The book echoed Thompson's insistence upon natural enemies that reproduced faster than the pests, and also his opinion that dramatically successful control could occur many years after introduction of the natural enemy. Sweetman also repeated an argument Thompson had made in favor of predators over parasites as biological control agents.[77]

Smith and his colleagues generally preferred to explore for

parasites, on the basis of their supposedly narrower host specificity. Thompson rejected this preference because of accumulating evidence that predators were more specific, and parasites less, than generally believed. Thompson claimed that predators served best because each individual ate more than one of the pest.[78]

The searching concept put this old biological control question in a new light. Reviewing Sweetman's text, Smith criticized the omission of any reference to density effects, searching ability, or Smith's argument for multiple introduction. Smith then made his case for parasites on the basis of searching ability. Parasites, he wrote, searched as winged adults and could therefore cover more ground than larvae, which were usually the most destructive stage in predacious insects. Also, Smith argued, parasites were best precisely because they needed only one host individual in order to survive and reproduce. A population of parasites could therefore survive on a smaller host population than could predators, just as a better searching species of parasite could survive at a smaller host density than a poorer searcher.[79]

S mith rarely communicated his theoretical musings on populations to California growers. By contrast, in earlier years, he had frequently cited the sequence theory as an authoritative, scientific basis for exploration and introduction of natural enemies. At that time, he had considered the theory well established, and he had no qualms about using it to build credibility with his public. In the 1930s, he told growers very little about host densities or searching ability. To that audience, Smith continued to speak of biological control as a matter of trial and error. As always, the objective was to find the native home of a pest, bring back its natural enemies, and hope that one of them controlled it.

Smith addressed his speculations on population dynamics to other scientists instead. He wished to justify biological control to those entomologists who doubted its value. He cautiously moved into ecological theory because his practical work obviously pertained to the concerns of ecologists. The change of audience was probably a long-term consequence of his shift from government service to the University of California. Working in an academic atmosphere, he aspired to be known as more than a pest control man. Information flowed from biological control practice to theory, but rarely did theory alter practice. Smith's conclusions on multiple parasitism and searching ability put old policies in a new light, but did not change them. Even most of Smith's own California colleagues understood little of his mathematical "hobby," except the basic logic of regulation. They were content to know that some scientific theory backed up what they had been doing all along.[80]

Although his mathematical work represented a step toward basic science, Smith emphasized its connection to practical problems. Near the end of his life, after others had turned population theory into a bitter debate over definitions and equations, he still saw his job as serving agriculture. Yet agricultural systems, even the California citrus orchards, were not always in the equilibrium to which Smith was committed by both practice and theory. Throughout the 1920s, citrus growers depended on mass releases of predacious ladybeetles to control citrophilus mealybug, because changing seasons disturbed the system every year. Left to work on their own, the beetles could not multiply fast enough after winter to overtake the mealybug before crop damage occurred.

In his theoretical papers of the 1930s, Smith acknowledged that physical factors could knock predator-prey interactions out of equilibrium. However, he considered agriculturally significant only those cases in which pests were at or near the maximum density a fairly constant environment would allow. Despite his experience with the mealybug, Smith focused on equilibrium level. He called the time necessary to reach it "a matter of little economic importance."[81]

Smith believed that most pests were in equilibrium, albeit at a density too high to suit farmers. He argued his position not from pest species, but from the vast majority of species that were not pests and never broke out in numbers noticeably larger than usual. Implicit in Smith's theory of density dependence was a basis for the heavy damage that insects inflicted on a crop in monoculture. This phenomenon, so familiar to American entomologists, could now be seen as a response to the plant's population density. Smith thought systems out of equilibrium to be exceptional. Practitioners of biological control aimed to remove such exceptions. As for citrophilus mealybug, parasites introduced from Australia had permanently reduced the population below pest status by the time Smith wrote on density dependence. In the end, he broadened his discussion to include insects such as grasshoppers, which characteristically broke out in large, destructive numbers and had not succumbed to biological control.[82]

Smith's basic views on populations paralleled those of Australian entomologist Alexander John Nicholson, who came to them not from practical concern over biological control but from consideration of evolutionary theory. Smith was no controversialist. He expressed his opinion as a tentative hypothesis, the beginning of discussion rather than a final, precise theory. Nicholson offered no such disclaimers. He hypothesized to greater detail—and drew a more heated response. A group of British and Australian ecologists, most of them well removed from practical work in biological control, debated density dependence and the relative importance of biotic and abiotic factors throughout the 1930s, 1940s, and 1950s. At the center of this dispute lay a personal exchange between Nicholson and one opponent who did work in biological control, W. Robin Thompson.

Nicholson arrived at density dependence by considering how natural selection operated in populations. He started with the Darwinian premise that populations remained in check despite their enormous powers of reproduction. Intraspecific competition allowed the most fit individuals to survive. Nicholson concluded that, in so doing, competition regulated population density by determining *how many* survived. Competition would intensify when the population increased, and relax when it decreased. Thus competition was density dependent. But Nicholson also realized that most species did not seem to exhaust their food supply, so some other density-dependent process must operate. Like Howard and Smith, Nicholson took population stability as given. Like them also, he concluded that natural enemies could be the balancing mechanism because they could increase with density, whereas climatic factors acted without regard to population size.

As Smith did at about the same time, Nicholson reasoned that searching ability, which he termed "power of discovery," would determine equilibrium prey density. Predator and prey were mutually regulating. Competition remained at the heart of the matter, because competition among the predators was the mechanism regulating both species. Trying to picture the outcomes of various combinations of hosts and parasites, Nicholson formed a series of hunches. For example, he reasoned that in some cases the addition of a second parasitic species could increase the population of the host. This argument directly contradicted that of Smith on multiple parasitism. Nicholson also believed that host and parasite populations would tend to overshoot equilibrium, not just in a steady oscillation but in one of increasing amplitude. He took his surprising conclusions to University of Sydney physicist Victor A. Bailey, who designed equations to match Nicholson's reasoning and produced the same outcomes.[83]

Nicholson published a long list of ecological conclusions from mathematics in 1933. By then, writers like Uvarov had questioned the very existence of the "balance of nature." Nicholson began by defending it. He then argued that only a density-dependent factor could achieve this balance. (Of course, he did not use the term "density-dependent," which was coined two years later.) That mechanism must be some form of "competition." Nicholson defined competition in a broad and nebulous manner. He repeated the arguments for balance and competition, with little change, for the rest of his career.

Nicholson believed that, from these postulates, he had deduced the detailed list of conclusions that poured forth from Bailey's mathematics—although the equations had been designed precisely to match Nicholson's arguments. This list included the implication that multiple introduction could impair biological control. Increasing oscillation would occur, due to the time lag between host increase and parasite response, but Nicholson argued that this took place in small areas within the larger environment. Emigration from areas of increase repopulated areas where the population had crashed. Thus Nicholson reconciled the inherently unstable system predicted by the mathematics to the overall balance that *must* exist.[84]

Smith welcomed Nicholson's paper, although it stole the thunder from his own. He saw Nicholson's work as a great collection of hypotheses waiting to be tested. Nicholson, happy to have Smith on his side, visited Riverside several times and later recalled their discussions as being among his most important scientific interactions. The Australian's treatment of increasing fluctuations especially caught Smith's eye. Nicholson suggested that hosts would appear in irregular, colonial distributions. Natural enemies would eventually find and destroy each colony, but in the meantime windblown escapees started new colonies. Nicholson initially gave no examples—his 1933 paper was remarkably free of references to actual insects—but Smith recognized the phenomenon right away. That most familiar of all biological control systems, cottony-cushion scale and vedalia beetle, behaved in exactly this fashion.[85] (Nicholson apparently never mentioned Smith to Bailey. The physicist was startled to come upon "a nest of supporters" during a visit to the Citrus Experiment Station in 1937.[86])

Robin Thompson was not as pleased as Smith. Thompson now realized that his own formulae had not worked. Old introductions still had not brought pests to the verge of extinction. He shifted completely, and became a vehement critic of attempts to equate mathematics and deductive reasoning with biological reality. Thompson believed that Nicholson and Bailey's equations contained too many assumptions that did not correspond to parasite biology. A theory without data could not stand. Furthermore, Thompson complained, the Australians had failed to credit him for anticipating some of their work.[87]

Thompson apparently still believed his own equations to be correct as far as was possible; their failure meant that no mathematics could suffice. In their place he constructed a theory of continuous environmental change. At any one time and place, he argued, environmental conditions either allowed infinite increase or drove a population toward extinction. Insects spread from favorable areas into unfavorable ones, which checked increase. Meanwhile new areas became favorable, allowing the species to persist. Complex ecological relationships among the European corn borer and its host plants, natural enemies, and physical environment presented just this picture to Thompson. He knew this insect best. He had spent nearly a decade trying to control the corn borer with natural enemies and had not succeeded.[88]

Thompson thus joined Uvarov in denying the existence of stable population balance regulated by biotic factors. He fired one written shot after another from the 1930s to the 1950s. Nicholson never failed to respond with the same arguments he had always used.[89]

The war of words between Thompson and Nicholson did not impinge much on their activities in biological control. Thompson continued to direct the movement of natural enemies into Commonwealth countries. Nicholson, who had previously done almost no research in pest control, took over as chief of the Division of Economic Entomology in Australia's Council for Scientific and Industrial Research (later Commonwealth Scientific and Industrial Research Organization) in 1933. Like the old agricultural bureaucracy in California, the CSIR had grown and sustained itself largely on the success of a single project—the biological control of prickly pear cactus—though this was actually accomplished by work done before the agency was formed. The CSIR naturally focused its efforts on its chief supporters, the cattle and wool industries. Its entomologists continued to work on biological control of range weeds, as well as various studies of the wool producers' nemesis, the sheep blowfly.

While acknowledging the unpredictability of individual cases, Nicholson did not stop his staff from making multiple introductions, despite his theoretical implication that they could impair control. His own research, on the sheep blowfly, dealt with intraspecific competition, not natural enemies. His rhetoric went in the same direction.[90] In both Britain and Australia, theoretical opinions—though deeply held—did not interfere with practice.

The absence of data to support either side allowed each to go on attacking the other. The first published attempt to verify density-dependent effects on searching and parasite reproduction came from Harry Smith's lab, in experiments he assigned to graduate student Paul DeBach in 1939. DeBach distributed housefly puparia and parasites in ten-gallon cans of barley, to measure the percentage parasitized at varying densities of each species. Rather than observing successive generations directly, DeBach and Smith counted the number of hosts parasitized in twenty-four hours and set up new cans with the resulting densities.

Practical biological control work beckoned, however—the smell emanating from the lab may also have played a part—and the experimenters were forced to stop after only seven "generations." Both densities dropped and then rose sharply in that time, but they did not complete even one period of the supposed oscillation. The data indicated a density response, but the authors admitted that their abbreviated experiment had proven little. They never returned to it. Critics dismissed it for its highly artificial nature as well as the failure to complete one oscillation, but almost no other efforts to test Smith and Nicholson's concepts appeared for years.[91] DeBach turned to experiments to verify biological control of pests, rather than density-dependent action.

Smith published little on population theory, while the war in print continued in England and Australia—and in Canada after World War II forced Thompson to move his operation there. Smith disliked controversy and ill will. Students and colleagues in California unanimously described their late leader as a kind gentleman who would search for a way to praise rather than criticize.[92] They were disappointed that he never wrote the biological control book he always talked about, but Smith could never seem to get to it.

The debate largely passed Smith by. Not a mathematician or really a theoretician of any stripe, he probably did not have many new ideas ready before other writers published similar ones. He presented a paper on insect population dynamics at the 1955 meeting of the Entomological Society of America, but never published it. By then he doubted that anyone could ever write equations successfully describing complex combinations of hosts and parasites of varying specificity. As he told his assembled colleagues, "In many ways the mathematicians have outdistanced the biologists in this field, and it might be well if a moratorium could be declared and the biologists be given time to catch up."[93]

Among the issues over which combative population ecologists argued in the 1940s and 1950s were the precise definitions of density dependent and density independent. Having adopted Smith's terms, ecologists disagreed on what the terms, or their author, meant. The squabblers rarely addressed agricultural insect control. New terms and theoretical twists filled the pages of scientific journals, but provided no guidance for experiment stations or for farmers trying to cope with pests. To Smith, this made the debate less relevant, but others pressed on.

The sharpest critics of Smith's terminology were two Australian insect ecologists, Herbert G. Andrewartha and L. Charles Birch. They directed their wrath primarily at their dogmatic countryman Nicholson, arguing that many populations were not regulated at all. Andrewartha and Birch opposed the distinction between density dependent and density independent, and assailed Smith's definitions on several counts in 1954.

Smith's graphs and the equation $m = aD + b$ from his 1935 paper treated the action of biotic factors as a linear function of density, and the proportion killed by physical factors as a constant (see Figure 4.1). He allowed that climatic factors could act in a density-dependent manner under certain circumstances. Such forces might, on occasion, kill only those insects unable to find refuge. The percentage killed would depend on the population size above the presumably fixed number of safe places.[94]

Andrewartha and Birch used Smith's statements about this effect of weather to claim that no such thing as density-independent mortality existed. It is reasonable to suppose, however, that Smith would have distinguished actual climatic effects from the density-dependent factor of competition for space, had the concept of ecological competition been precisely defined by 1935. He did express this distinction in later years.[95]

Smith possessed a limited knowledge of mathematics. He lacked an understanding of statistical variation, which would have been proportionally greater in small populations than in large ones. More important, he did not intend to describe a precise mathematical relationship. He represented density dependence schematically. He may have originally expected the relationship to approximate a line, but he did not claim to have proven it. He inserted the disclaimer, "This equation is as representative of the real facts of nature *as can be expected when mathematics is applied to biological problems*" (italics added).[96] He did not believe that mathematical theory could yet describe nature precisely. Nevertheless, critics such as Andrewartha and Birch took him to mean that *a* and *b* were constants.

Of course Smith knew that physical factors varied—weather, for example. The "constant" *b* meant only that a *particular* set of physical conditions would kill the same proportion of insects regardless of density. The notion of a constant in *all* conditions was a straw man that did not match the common usage of "density independent." Likewise, Smith made it clear that the environmental conditions in which a biotic agency acted determined the magnitude of the density response. The more intense the response, the lower the equilibrium density of the host. In any particular set of conditions, however, the relationship did appear to be linear in his schematic representation. The rate at which mortality increased with density appeared as the slope (*a*) of a line.[97]

Whether the effect was in a general sense linear or not, the steady equilibrium Smith pictured in 1935 depended on the immediacy of the response. By 1939, after considering the oscillations inherent in Nicholson's theory, Smith acknowledged that the response of natural enemies would show a time lag. Parasite reproduction required some time after the host became more abundant. Oxford professor George C. Varley, who had spent a postdoctoral year with Smith in the 1930s, led a movement two decades later to add a new term. Varley called these factors "delayed density dependent" to distinguish them from truly "density-dependent" factors. As another Briton quickly responded, Varley was denying Smith's common usage, which allowed for lags, in favor of an original definition more precise than Smith had intended. Smith, who died before this exchange took place, did believe that natural enemies were legitimately called density dependent because they responded to increasing density, even if the response actually occurred a generation later when host density might no longer be increasing.[98]

Posturing over definitions was not Smith's concern. He earned a place in the history of population ecology because ecologists grappled with his terms for decades. But he aimed primarily to provide a scientific justification for biological control as practiced in California. Practical justification he already had, in the conspicuous and popular success of introduced natural enemies against a small number of citrus insects.

Before the 1950s, Smith's staff largely stayed out of the population debates. He, his assistants, and his students did, however, pursue other biological questions that more directly affected the course of biological control work. Some of these, to be discussed in the next chapter, became more pressing as the success rate failed to climb during the 1930s.

Smith's approach ultimately remained practical. When he arrived in Sacramento in 1913, he channeled California's early enthusiasm for biological control into a realistic program. The program still traded on its popularity with growers, especially citrus growers, but he reminded them not to expect too much too soon. He helped make them active participants, when simple importations fell short of complete control but mass production proved effective. Smith secured a unique institutional identity for biological control in the transfer to the University of California, and established himself as an academic scientist without forsaking his practical mission.

Successes were few, but biological control prospered as a field of study in California. Smith earned the good graces of the U.S. Department of Agriculture and preserved California's right to pursue the biological method in ways the USDA itself was not prepared to do. His cautious approach and pleasant personality won the support of agriculture, academia, and government. Yet the program remained closely bound to one agricultural sector. An environmentalist concerned about the extensive use of pesticides half a century later might look back and ask why the biological control group did not do more. It is quite possible, however, that without such close ties to citrus in California—and to sugar cane in Hawaii—the method would not have survived at all.

Systematics, Evolution,
and Applied Entomology

The 1920s ended with the second solid success of biological control in California, four decades after the first. The complete and permanent subjugation of citrophilus mealybug by deliberately introduced parasites cemented the support given to biological control by the citrus industry. Growers thus reaped their first practical reward since the University of California took over biological control from state government and provided an academic setting.

In that setting, entomologists found greater opportunities for scientific research, but the primary objective remained the same. As we have seen, developments in the 1920s helped propel California practitioners into theoretical ecology. In the years that followed, efforts to control pests led to questions of systematics, evolution, development, and genetics. Entomologists seeking new parasites abroad could no longer tell which pests were which, let alone untangle the relationships among natural enemies.

Differences that could not be detected by old methods proved decisive in the orange groves. Most alarming to growers, scale insects evolved resistance to the most common insecticidal treatment in citrus. This problem would strike other sectors of agriculture after World War II. In California citrus, resistance was already a serious concern that increased pressure for a permanent, biological solution. These frustrating practical problems raised enticing scientific issues as well.

As head of the Division of Beneficial Insect Investigations at Riverside, Harry S. Smith still put the financial interests of growers above pure-science objectives. It became increasingly clear, however, that research was not only possible alongside the practical mission, but sometimes essential to achieving it. State government service had allowed Smith and his associates little time for scientific investigation. The university, which they joined in 1923, encouraged—and, increasingly, demanded—original research for professional advancement. Such obviously relevant topics as parasite behavior and development were well within reach of biological control workers. These entomologists were less well placed to study systematics and evolution, however, and professionally distant from areas such as genetics, at a time when the developing evolutionary synthesis attracted many non-economic biologists.

Both cottony-cushion scale—by now almost the stuff of legend—and citrophilus mealybug, newly subdued as the decade began, especially affected citrus. Smith kept citrus insects, especially black scale and California red scale, at top priority through the 1930s. In return, the industry offered moral and tangible support for research. The entomologists continued to encourage that sponsorship and work closely with the growers. In name a statewide entity, Smith's division nevertheless maintained its emphasis on citrus from headquarters at the Citrus Experiment Station. The staff functioned as an integrated team with a common goal. The industry's needs grew more intense as the Great Depression and uncontrolled expansion of supply drove fruit prices down. Growers needed to reduce costs to stay in business. Chemical means of insect control stood among the largest expenses.

Specific obstacles to application led to fundamental advances that helped entomologists to introduce more parasites into California, but without stunning successes to match the first two. It was still necessary to warn growers against expecting miracles. The groundwork for future success was laid, however, both in the expanding research agenda and in the training of a new generation of biological control scientists. Only in California could students obtain significant graduate education in biological control.

Smith rose to the pinnacle of his profession by the end of the decade, when he was elected president of the American Association of Economic Entomologists. Although few Americans outside California made much effort in biological control, entomologists in the United States and, indeed, around the world acknowledged Riverside as the leading center for natural enemy studies. Smith probably owed his election to his assistants' accomplishments as well as his own.

He used the occasion to issue a thorough but polite warning that insecticides could sow the seeds of their own demise. Years later, when environmentalists attacked insecticides for their possible effects on wildlife and human health, some entomologists were changing their approach for an entirely different reason: The chemicals weren't doing the job on the pests.[1] Smith had been right. Striving to be part of exciting movements in biology, especially in evolutionary theory, Smith brought his own experience from citrus to bear on the development of resistance to chemicals. Even then, he retained the economic outlook of his colleagues in chemical control, and considered himself a servant of the farmer. He directed the work in Riverside from this perspective.

The Division of Beneficial Insect Investigations continued to operate as a team during the Depression years. Increasingly removed from the insects themselves, Harry Smith still supervised the division as a team captain. He selected biological control projects—with considerable input from the citrus industry—and planned foreign exploration for natural enemies. He appeared at growers' meetings, wrote articles for citrus magazines, and on one occasion even went on radio. He gave prognoses on the most important citrus insect problems, but reserved his mathematical ruminations on population dynamics for another audience, his fellow scientists.

The men Smith appointed to the key positions of foreign explorer and insectary manager faced taxonomic problems that helped arouse his interest in evolutionary theory. He had hired a series of wandering entomologists since 1914. Most remained on the job only a year or two without distinguishing themselves. Then, in 1927, on his first time out, Harold Compere neatly solved the citrophilus mealybug problem. He became the permanent explorer for the division, roving worldwide as his father, George Compere, had done in the early years of the century. Father and son each lacked formal training. Instead of boasting and bickering, however, the younger Compere educated himself on the parasitic Hymenoptera between trips. Although he was unable to match his initial triumph, Harold Compere's travel tales helped to keep the growers interested in biological control.

Many of the insects shipped to California by earlier explorers died en route. Others arrived alive only to fail in laboratory culture. Each species required somewhat different conditions for rearing in the lab. The trial-and-error process of establishing those conditions wasted many a foreign shipment. The last crucial role on the team at Riverside was a laboratory entomologist to screen shipments in quarantine, solve rearing problems quickly, and keep cultures of entomophagous insects alive long enough to try in the orchards. Smith had hired Philip Timberlake for this purpose in 1924, but Timberlake would not budge from his interest in wild bees. The job was filled at last with the arrival of Stanley E. Flanders in 1929. While on the staff, Flanders also became Smith's first doctoral student.

Flanders came on board initially for mass production of beneficial insects. Annual releases of ladybeetles against citrophilus mealybug had been rendered unnecessary, but mass production remained an interest of the biological control group. The son of a San Diego County grower originally from New England, Flanders was born to citrus. He moved through various jobs before completing a bachelor's degree at Berkeley in zoology, with an emphasis in entomology, at the age of 29. The Saticoy Walnut Growers Association, a cooperative in Ventura County, hired him out of college in 1923 to deal with the codling moth (*Carcocapsa pomonella*). The widespread apple pest had recently taken to walnuts.

Investigating the possible value of parasites, Flanders soon made contact with Harry Smith, and at Smith's suggestion began to study egg parasites of the genus *Trichogramma*. The tiny wasps were already common in the area, but Flanders thought he might be able to control codling moth with regular releases of *Trichogramma*. Apparently neither Smith nor Flanders knew at the time that artificial production of these parasites had been tried in Russia a decade before and had been suggested in England as early as 1895. Smith noted that any biological solution to codling moth would be a boon because of the alarm raised in Great Britain in the mid-1920s over arsenical residues on American apples and pears.[2]

With the financial backing of the walnut growers, Flanders developed a mass rearing technique using another species of moth as a host. By the time he left Saticoy in 1929, production had reached 600,000 eggs per day. His work caught the attention of entomologists all over the United States and in other countries, since *Trichogramma* could parasitize almost any undesirable species of moth. Flanders was besieged with inquiries. His technique became the standard of a *Trichogramma* industry aimed at everything from peaches in Colorado to sugar cane in Louisiana, although he cautioned that he had no evidence thus far as to practical value. He came to Riverside, broadened his research on *Trichogramma*, and completed a doctoral thesis on the job.[3]

The claims others were making for the little egg parasites disturbed Smith as early as 1928, before he hired Flanders. Smith had long tried to eliminate exaggeration from biological control, but now that specter of earlier times had returned. In particular, Warren E. Hinds of the Louisiana Experiment Station threw caution to the wind and recommended artificial distribution of *Trichogramma* as a practical measure against sugar cane pests. Closer to home, Austin Winfield Morrill set up a private *Trichogramma* business at the Orange County Insectary, in space no longer needed for mealybug predators. Morrill's California Insectaries thrived and soon moved to its own quarters.

Flanders helped the outfit get started, then watched in dismay as Morrill promised benefits from the purchase of *Trichogramma* for a variety of field crops. To dissociate themselves from the venture, Smith and Flanders published a stinging criticism in 1931. Labeling *Trichogramma* a fad, they warned that its promoters could not yet prove commercial releases increased parasitism in the field. Morrill and Hinds continued to press their claims. By the end of the decade, however, USDA entomologists rejected artificial use of *Trichogram-*

ma, finding no value in Louisiana sugar cane or other crops. The fad faded until the 1950s.[4]

Easily reared in the laboratory, *Trichogramma* remained a useful organism for study. Flanders examined the effects of temperature and other factors on the wasps' development. He also investigated possible races—or species—of *Trichogramma* that differed by color or preferred habitat but not by any known, consistent, morphological characteristics. His interest in forms that traditional museum taxonomists could not distinguish may have been triggered by the presence of William H. Thorpe, a British postdoctoral student completing a two-year stay in Smith's lab when Flanders moved to Riverside. Thorpe's primary concern at the time was the evolution of races within species of insects. Even without Thorpe's influence, frustration from working with the taxonomically chaotic genus *Trichogramma* might have been enough to convince Flanders that developmental circumstances could cause intraspecific morphological differences greater than those used to distinguish between species.[5]

Flanders believed that the common *Trichogramma* did comprise several species, but that these could be distinguished only by biological differences such as habitat selection and response to temperature. Others working on the genus failed to find a morphological basis for separating species, but were loath to accept the broader implications Flanders drew. He concluded that, in some cases, no reliable morphological differences would *ever* be found.[6]

This period found taxonomists in a quandary. With the general acceptance of evolution since the time of Charles Darwin, biologists increasingly recognized that morphological distinction was not the essence of species. Systematists gradually adopted what Ernst Mayr has called "population thinking," recognizing that species were populations in which variation was to be expected, and that no individual "type" could truly represent the whole species. The closest thing to an ultimate criterion for determining if two forms were the same species or not was whether they could successfully interbreed. Nevertheless taxonomists, particularly insect taxonomists located in museums, still relied on dead specimens and thus on morphology alone. A systematist would have to judge whether the morphological variants within two populations intergraded too much for them to be called separate species. It was hoped that species determined in this way would satisfy biological criteria. The possibility that true species could be utterly indistinguishable was still more than some workers could take.[7]

It would be useful to know what proportion of systematists in various biological specialties were really affected, in thinking and in practice, by developments in evolutionary theory prior to the 1930s. Certainly an influential school of Darwinian systematics arose in entomology under the leadership of John Henry Comstock at Cornell from the 1880s onward. Comstock emphasized functional and evolutionary significance in deciding which morphological traits indicated natural systematic relationships. He was primarily concerned, however, with relationships among higher taxa, rather than questions of splitting or lumping species.[8] We do not yet know how deeply Comstock's most devoted followers, let alone the rank and file of insect taxonomists, pondered the species concept itself.

No evidence ties Flanders to the more progressive thinking in systematics or evolutionary theory in the late 1920s or early 1930s. He apparently made no crossbreeding tests among the forms he suspected of being distinct species, although at least one of his correspondents working on *Trichogramma* did.[9] Yet he argued, as few had up to that time, for what Mayr would define in 1942 as "sibling species."

Mayr attacked the artificiality of a strictly morphological species concept, and noted that entomologists clung to it especially tightly. By the 1930s, they had encountered an increasing number of groups separable by biological characteristics but not by morphological traits. In another case of applied work, contemporary with Flanders's *Trichogramma* studies, medical entomologists discovered sibling species in *Anopheles* mosquitoes. The forms differed in their ability to transmit malaria.

At that time, most entomologists still preferred to call such groups races rather than distinct species. In museum settings, using only dead material and having to deal with an enormous number of species, most insect taxonomists could not distinguish and name sibling species. As Mayr pointed out, it is sometimes difficult to tell whether Thorpe and others were referring to races within a species or to truly distinct species that happened to look alike. This is also true of Flanders's early papers, but he clearly moved toward the sibling species concept during the 1930s.[10]

Among evolutionary theorists, the question of whether reproductively isolated, but morphologically indistinguishable, forms could exist gave way to the more fundamental matter of how one species actually split into two. The issues posed problems of nomenclature even for taxonomists less theoretically inclined. Until Mayr provided a satisfactory term, biologists who recognized that sibling species existed disagreed over what to call them. Were they "species" or merely "races"? Should standard binomial designations be given to forms the average worker could not identify from morphological literature and dead specimens? If not, then how should such forms be designated? Biological control work would require answers to these questions raised by *Trichogramma*. The most subtle differences among either pests or natural enemies could derail a project.

Trichogramma also stimulated basic research in ecology. Biologists used the tiny wasps to investigate the process by which adult female parasites locate and select, or reject, hosts in which to lay eggs. At the Farnham Royal, England, headquarters of the Imperial Bureau of Biological Control (later Commonwealth Institute of Biological Control) under W. Robin Thompson, George Salt demonstrated that the females could distinguish between parasitized and unparasitized moth eggs. He found that this ability to reduce the incidence of superparasitism depended on the detection of a chemical the females left behind. Salt's experiments helped launch the study of what would later be called pheromones. Other work at the British lab concerned species being considered for introduction to various parts of the Commonwealth. Thompson encouraged such studies for their possible contribution to biological control as well as basic science. He even pleaded for people to come and use material available at Farnham Royal for research that Depression budget cutbacks prevented him from supporting directly.[11]

On both sides of the Atlantic, then, entomologists interested in biological control began to ask basic questions beyond the location of predators and parasites for a given pest. Low success rates belied earlier claims that natural enemies would provide a quick and easy solution to most insect problems. It was hoped that data on ecological characteristics might help guide the search for and selection of biological control agents, or facilitate periodic release programs such as those using *Trichogramma*. Perhaps, with enough information, entomologists could determine which pests were least likely to succumb to natural enemies, so that waste of time and resources might be avoided. Advances in systematics might lead to re-exploration of previously covered territory, to find species that had been missed or discarded without trial because they were thought synonymous with others. In the years to come, the Californians would argue repeatedly that the dearth of such information delayed progress in applied biological control. Practical missions motivated fundamental research.

In the case of *Trichogramma*, application gained little in the 1930s. Flanders and the rest lacked any dependable basis even to tell which species they had in hand at any one time. Consistent morphological characters to untangle the genus were discovered only during its resurgence as a commercial product in the 1950s and 1960s. Entomologists then found that, in many cases, the species parasitizing large numbers of pest eggs late in the season was not the one farmers bought from private insectaries. Therefore, no value could be attributed to the commercial releases, except that the peace of mind from doing *something* may have restrained farmers from spraying chemicals that would kill beneficial insects already present.[12]

Flanders made more direct impact on the California biological control program with his efforts to culture new parasites sent from foreign countries. A species of *Coccophagus* parasitic on black scale (*Saissetia oleae*) arrived from Brazil in 1935, but Flanders found it difficult to keep the *Coccophagus* going in the lab because of a shortage of males. He was employed to solve just this sort of problem. Lopsided sex ratios in laboratory populations commonly obstructed the establishment of new species. Although the South American black scale parasite never amounted to much as a control agent, it led to the most important discovery of Flanders's career.

As in other Hymenoptera, unfertilized eggs of the *Coccophagus* developed into males, and fertilized eggs into females. Flanders made sense of observations dating back to 1913 that certain species of the family Aphelinidae, generally considered primary parasites, sometimes acted as hyperparasites on closely related species or even their own. Flanders noticed that all emerging hyperparasitic *Coccophagus* were male, the primary parasites female. This had been observed in other aphelinids, but the only explanation anyone had suggested was that hyperparasitic development caused maleness.

When Flanders provided only parasitized hosts to mated adult *Coccophagus* females, the offspring failed to develop. The offspring of virgin females could not develop in unparasitized hosts. Flanders concluded that in these insects the adult female, once mated, could lay only fertilized—that is, female—eggs, which became primary parasites. Male eggs, laid only by unmated females, were obligate hyperparasites on developing female larvae.

This phenomenon, widespread in Aphelinidae, could explain past failures to maintain populations in the lab or to establish species in the field. Mated females had to find unparasitized host scale. Virgin female offspring, upon emerging, must lay eggs on other females still in larval stages in other hosts. Propagation of the species required correct timing: When the males emerged, there still had to be some females around for them to mate in order to produce the next generation. In some species, the choice of parasitized or unparasitized scale depended on whether the female had mated, while in others oviposition behavior was just a matter of chance. Especially in the case of males that could attack only their own species, this sequence of events might be very difficult to recreate in a small population. In the wake of Flanders's discovery, biological control workers at least knew what they needed and could keep more species of this important parasitic family alive long enough for trial in the orchards.[13]

Flanders's chief concern was maintaining parasites in culture so that they could be used in field trials, in service of agriculture. These efforts, taken together with his advancement of scientific understanding of parasite development, provide another example of the difficulty of classifying the work at Riverside as strictly science or technology.

Pest control is technology, indeed. Flanders certainly acted as an

engineer, tinkering with culture conditions. But economic entomologists would have called themselves scientists rather than engineers.
They identified with scientific knowledge and scientific organizations.
Sibling species and male hyperparasitism—and Harry Smith's own pet
interest, population dynamics—were scientific matters. Yet agricultural
scientists studied them out of technological need. Flanders followed a
complete cycle: He encountered a technological difficulty, made a
basic science discovery, and applied it directly to the original problem.
Something akin to the model of mirror-image twins may hold if the
twins are scientific and technological activities, rather than separate
communities.

Industrial support of pest control technology was never more
important than during the Great Depression. Tales of entomological exploration piqued the interest of growers. The biological
control program owed its existence to their patronage, which Smith
and his staff continued to cultivate. Explorers' voyages made good
copy. Harold Compere generally stayed abroad for a year or so at a
time. On returning home, he was in demand as a speaker and writer.
Status reports and stories of the road splashed across the pages of the
California Citrograph, the citrus monthly dominated by the California
Fruit Growers Exchange (Sunkist). Smith, and occasionally his
assistants, spoke at pest control clinics organized by county extension
agents, at citrus institutes of the annual National Orange Show in San
Bernardino, and at meetings of the Southern California Entomological
Club and the Lemon Men's Club, a social organization of growers.
Smith went on the air in a 1932 radio address entitled "Insect Friends
of the California Farmer."[14] In each venue, entomologists kept
biological control before the customers.

Industry support continued, although by the 1930s natural
enemies had completely controlled only two pests in half a century.
Growers welcomed anything that would cut their production costs.
Prices for oranges, still something of a specialty food, fell even faster

than those for other farm products at the onset of the Great Depression. Orange acreage continued to increase. The *Citrograph* advised against cutting corners in pest control, but growers needed all the help they could get. As if the industry did not already exert enough influence over the University of California, Sunkist president Charles C. Teague joined the university board of regents in 1930. With such backing, Smith obtained a new $30,000 insectary in 1931. The new facility was more insect-tight and better suited to parasite production than any built previously.[15]

The enthusiasm of one agricultural industry in California, and the consequent focus on that crop by the biological control team, in some ways paralleled developments in the U.S. territory of Hawaii and the British colony of Fiji. In Hawaii, the sugar planters themselves employed a staff of entomologists dedicated to the use of natural enemies, with the cooperation of the Territorial Board of Agriculture and Forestry. The board added pineapple work as that industry also arose in an oligopolistic plantation system. Planters believed in biological control, paid handsomely for its pursuit, and exerted political pressure to sustain it. (See Chapter 3.)

Similarly, coconut plantations supported the use of predators and parasites in the Fiji Islands. There also, a team concentrated on a single crop and put together a string of successes against insects and weeds in the 1920s and 1930s. Biological control found favor because of the great expense of pesticides relative to the value of the crop, and perhaps also because the haughty colonists considered native laborers incapable of implementing chemical control on the plantations. British entomologists working for the colonial government handled both foreign exploration and domestic phases of the work.[16]

Success rates may have been somewhat exaggerated in Hawaii, especially before the University of Hawaii entered the field on behalf of the pineapple industry in the 1930s. Government and Hawaiian Sugar Planters' Association entomologists tended to count each establishment of a new natural enemy as a success, regardless of the actual effect on crop damage. Accurate or not, the record in Hawaii and Fiji suggested to some that islands were inherently better suited to biological control than were continental areas.

As first expressed in the 1920s, the idea meant simply that introduced beneficial species would be unlikely to encounter effective natural enemies of their own, particularly hyperparasites, among the limited fauna of a place as isolated as Hawaii. British entomologist Augustus D. Imms transformed this notion into a hypothesis that distinct boundaries, rather than distance, gave islands their advantage. Success in California challenged his pessimistic view of continental areas. Imms deflected the challenge by labeling California's agricultural region an ecological island. Others argued that the tropical islands owed their success to the absence of sharp seasonal changes. In such an environment, pest and parasite alike could live in stable populations. All stages of each would be present at virtually any time of year.[17]

The Californians dismissed the island hypothesis altogether and minimized the importance of seasons, pointing to successes in such seasonably cold climes as Canada and New Zealand. California, Fiji, and Hawaii led the way in results because they led the way in effort.[18] The *kind* of effort—an integrated, team approach with governmental and industrial assistance—made biological control pay. Concentration on one crop at a time, while methods were being worked out, made the best of limited resources. Close contact with an audience of farmers willing to believe in biological control increased the chances of success in the present, and of continued funding in the future.

Support for research in both biological and chemical control of insects was so strong in the California citrus industry that Harry Smith's Division of Beneficial Insect Investigations and Henry J. Quayle's Division of Entomology dwarfed the rest of the Citrus Experiment Station staff. The entomologists' status sparked jealousy among other divisions such as plant pathology.[19] Divisions worked together, however, to defend one of the basic assumptions of the industry the station served.

Citrus entomologists and plant pathologists agreed that the best way to deal with insects and diseases of agricultural plants was to keep the pests from reaching California in the first place. Critics characterized plant quarantine laws as disguised tariff measures, intended to keep competitive produce out of domestic markets. This no doubt was the primary purpose of some international quarantine laws, but California growers' fears of exotic insects grew out of very real experiences. Virtually every pest of citrus was foreign. A foreign disease such as citrus canker might create an even greater disaster than a new insect.

Dissenters included Charles W. Woodworth of Berkeley and, most vehement, Gordon F. Ferris of Stanford. Woodworth and Ferris argued that border and port inspections could not possibly prevent new species from entering California. Eradication efforts such as the expensive campaign against the Mediterranean fruit fly in Florida in 1929 were doomed if the pest could find suitable conditions in which to live. Smith replied that, while inspections obviously would not catch every insect, they caught enough to prevent or delay establishment of new pests. Although he complained about the time taken from his other work, Smith chaired a committee of scientists that produced a comprehensive argument for the effectiveness and justification of plant quarantines.[20]

Growers were sensistive to anything that threatened their livelihood. Some were staring at bankruptcy during the Depression. The collapse of consumer buying power countered the marketing practices by which the California Fruit Growers Exchange had increased the demand for oranges and lemons. Efforts to hold rival citrus groups in California and Arizona to marketing agreements during the emergency were shaky at best. Even 90 percent participation failed to bolster prices. Industry leaders finally admitted that spreading their fruit carefully around the national market could no longer hold prices high enough. Demand leveled off, while supply continued to increase as more trees came into production.

Growers would have to withhold some fruit. However, the organized industry was powerless to force all independents to go along even after the state legislature passed a bill authorizing mandatory proration of shipments in 1933. Although members of Sunkist produced nearly 75 percent of citrus shipped from California, nonmembers were given sufficient representation on the prorate-setting board to block marketing orders, which they eventually did. Attempts to bring less solidly organized Florida and Texas growers into the fold failed completely. Under any prorate, smaller operators would suffer the most.[21]

The only point on which all could agree was that a reduction in costs, such as those for insect control, would help. Scale insects, particularly California red scale (*Aonidiella aurantii*), topped the list of problems for both groups of entomologists at the Citrus Experiment Station. The usual treatment, tent fumigation with cyanide gas, was becoming less effective in several citrus areas of southern California. Henry Quayle had been writing about the phenomenon since 1916. Most economic entomologists across the United States questioned the evolution of resistance to an insecticide, although a few other instances had been reported.[22] In Riverside there was little doubt. Not only red scale, but also black and citricola scales were becoming resistant to the usual dosage of cyanide. Heavy infestations of red scale could kill the trees, but growers faced a more subtle problem. Stringent cosmetic standards in the industry meant that even a few red scale on the rind, though harmless to the fruit inside, could make the difference between profit and loss. Something had to be done.

S mith may have been tied to his desk more than he liked, but his subordinates roamed far and wide. Basic biological questions slowly emerged from Harold Compere's efforts to find new parasites and "bring 'em back alive." Much of the frustration that could later, following Flanders's discovery, be blamed on male hyperparasitism surrounded the long campaign to control black scale. A number of parasites sent from South Africa in the 1920s were considered failures. Following his smashing success against citrophilus mealybug, Compere sailed for Eritrea, where black scale parasites had once been reported. Uncertain whether the pest actually existed in that part of Africa, and wanting an easy way to get parasites back home if he did find any, Compere brought scale from California on small sago palms. He used this host plant to avoid strict quarantine restrictions forbidding importation of citrus from abroad. The near destruction of his plants and insects en route by the hot sun on the ship's deck became the first of many pitfalls he would use through the years to dramatize his collecting expeditions.

Once he reached Africa, Compere lived almost as a country gentleman, complete with servant. He found a few potentially valuable parasites, let them into the cages to attack his remaining black scale, and packed up for home. Then, while he tried to make connections via the Suez Canal, his plants were again exposed to the tropical sun, and everything was lost. Compere vowed to leave exploration to someone else, but he was on the road again before long.[23]

The original plan called for Flanders to take a share of foreign exploration, so in 1931 he traveled to Australia in search of enemies of citricola scale (*Coccus pseudomagnoliarum*). This soft scale insect especially troubled growers in the hotter, drier inland citrus districts. Its place of origin (now thought to be Asia) was still uncertain, but Australia seemed a safer place to start looking than war-torn China. Flanders found no citricola scale. He did add to the lore of dangerous travel, breaking his collarbone in an automobile accident. He took no more foreign trips for more than twenty years, mainly because he suffered so badly from seasickness.[24]

Although the citricola scale problem remained unsolved, the latest Australia voyage helped the boss in other ways. Flanders made the first contact between the Californians and Alexander John Nicholson, who was independently developing ideas on population dynamics similar to Smith's. Flanders also reported back to Smith on the excellent progress the Australians had made in biological control of prickly pear cactus. By this time, Smith had taken a serious interest in weed control. He wanted to know whether Australia was making any headway against another range weed, St. John's-wort (*Hypericum perforatum*). California cattle ranchers were beginning to lobby the university for introduction of insects that fed on the European plant, known locally as Klamath weed. Another decade would pass, however, before federal authorities permitted Smith to attempt biological control of weeds.[25] (See Chapter 6.)

Back in California, Flanders settled into his primary duties as quarantine and insectary supervisor on the team Smith built in

Riverside. Flanders received shipments of foreign insects, and separated beneficial species from potential pests. Access to the quarantine room was strictly limited to reduce the chances of a new pest escaping the lab. Flanders also attacked puzzles such as that leading to his discovery of obligate male hyperparasitism. Among common sources of frustration were parasites that would not attack the same insects in California as they were reported to parasitize back in China or Australia. By this time, Smith and Flanders believed that they had an explorer knowledgeable enough to identify scale insects and parasites in the field, yet some species seemed to change their habits upon crossing the ocean. Such obstacles to the application of biological control led Smith's group into taxonomic and parasitological research often unforeseen.

California red scale, in particular, received increasing attention and posed one riddle after another. One of the most perplexing concerned a parasite known as *Comperiella bifasciata*, often reported on red scale in the Orient and shipped to California without becoming established. From material collected by George Compere, Leland O. Howard described a new genus and species in 1906. Howard, chief entomologist of the U.S. Department of Agriculture, considered much of the elder Compere's work useless or even dangerous. Nevertheless, Howard named the new genus for the indomitable explorer. More *C. bifasciata* arrived from Japan in 1916, 1922, and 1925. Explorers reported finding the parasite on California red scale, but eggs laid in red scale in the lab failed to develop.[26]

The last sending, however, did get *C. bifasciata* established in California on another scale insect, a minor pest of ornamental plants. California entomologists at that time considered the latter scale, a species of *Chrysomphalus*, to be closely related to California red. They did not yet accept the separate genus *Aonidiella*. In 1931, Smith suggested trying the parasite on yellow scale, *Aonidiella citrina*, a significant citrus pest in its own right and the closest known relative of *aurantii*. The parasites developed without difficulty and soon cleared residential citrus trees of yellow scale in the southern counties, although the scale remained a problem in orchards. The ability of *C. bifasciata* to attack scales less closely related than yellow was to red, without attacking red, left Smith and his assistants scratching their heads.[27]

Still, the fact that *Comperiella* parasitized yellow scale revived hopes that Asia might yet yield valuable parasites of California red scale. As early as 1913, Smith had suspected that explorers reporting *Comperiella* on red scale might have seen a race or even species of scale morphologically indistinguishable from *Aonidiella aurantii*, but vulnerable to parasites to which true *A. aurantii* was immune. The distinction between *aurantii* and *citrina* had been in dispute for forty years, because no consistent morphological character had been found to separate them. *A. citrina* was lighter in color under some conditions, but not all. It attacked only the fruit and leaves of citrus in southern California, whereas *aurantii* thrived on the bark as well. Systematists in the 1890s would not generally accept such ecological characters to separate species. Charles Valentine Riley, then the most prominent American entomologist and a key player in the beginning of biological control, refused to separate red and yellow scale without structural characters. In California, though, most entomologists went along with the experience of growers that these two pests were distinct despite their similarity.[28]

The behavior of *Comperiella bifasciata* led Smith to suppose that explorers prior to the 1930s failed to tell red and yellow scale apart in the Orient. This confusion would explain repeated shipments of parasites reported from red but able to attack only yellow. Smith began to doubt that any effective parasites existed for red scale. Another possibility, however, was that the parasite *Comperiella* existed in two forms, and that the explorers had indeed observed it on red scale yet happened to ship only the form from yellow scale.

On that hope, Harold Compere sailed for India and China in 1932. In order to test any parasites he might find, he carried true California red scale on sago palms. If entomologists could not identify the scales, then, as Compere wrote to Smith, "By their parasites ye shall know them." Despite all his careful preparation and study before leaving, Compere promptly misidentified yellow scale as red in India. Then, in a hurried telegram, he reported a new species of *Comperiella* to be an effective control. After the parasites failed to develop in the red scale brought from California, Compere realized that not only had he found the wrong species of pest, but the parasite he thought new was the same old *C. bifasciata*. Trusting only in the scale on his palms, he found no parasites that could live on it in China. Pessimism reigned again in red scale control.[29]

Growers insisted that Smith keep trying. Some entomologists believed that the genus *Chrysomphalus*, to which they still assigned red scale at the time, had originated in Central or South America. Smith told growers that only those regions remained potential sources of natural enemies. He warned that the chances were poor, but growers insisted that even the dimmest prospects must be investigated. In spite of marketing disputes, Sunkist and four smaller, rival cooperatives joined two citrus shipping companies in contributing more than $7,700 to send Harold Compere to South America. The Lemon Men's Club added $1,000 so his wife, Joan, could join him as she had on earlier trips. The customers had spoken. Smith acquiesced and dispatched Compere on a seemingly hopeless mission in 1934.[30]

Smith feared that South American bureaucrats would snarl his explorer in red tape without full documentation by the federal government. Smith arranged with the U.S. Bureau of Entomology and Plant Quarantine for Compere to travel as an official agent of the USDA. In return, Smith promised that Compere would keep an eye out for insects that interested the bureau. Compere also looked for enemies of pineapple pests, since Californian and Hawaiian biological control workers frequently needed each other's cooperation in relaying shipments. Smith made it clear that most of the explorer's time must go to citrus because the growers had paid for the trip. A vegetable insect became a legitimate target when it moved into cover crops used in California citrus orchards—"What a break for us!" the boss wrote to Compere before the latter came home in 1935.[31]

Compere returned without much good news about red scale. In Brazil, the scale infested roses heavily but was rare in adjacent citrus. Compere could not determine why. Natural enemies appeared to have nothing to do with it. Trying to find some return on the citrus growers' investment, he discovered a *Coccophagus* species parasitic on black scale. At the receiving end, Smith and Flanders were briefly excited for black scale control. Their hope faded quickly, but the parasite led to Flanders's studies of male hyperparasitism.[32]

Flanders's discovery reawakened interest in black scale, many parasites of which had been shipped from Africa over the years without being established in California. Africa also stirred new hope for red scale, which remained at the fore. Entomologists in South Africa sent specimens for identification, including a new species of red scale parasite that Compere named *Habrolepis rouxi*. More than half of the growers' fund for the South America expedition remained. Smith, still pessimistic about red scale control, sent Compere to Africa in 1936 to bring *Habrolepis* back alive and to check on black scale parasites as well.

Compere received ready cooperation in South Africa. A government entomologist who had studied red and yellow scales at Riverside prepared a laboratory in Capetown for scale and parasite rearing. The first sight of *Habrolepis* attacking red scale so impressed Compere that he forecast the scale's doom in California. He hoped the only reason for high scale populations in South Africa was that the parasite had arrived there only recently and had not yet spread throughout the citrus areas. "Can you imagine how I will feel if *Habrolepis* proves to be another false alarm?" he wrote. It was. *H. rouxi* had been in South Africa long enough to help if it was ever going to. Although it became established in California, the parasite avoided citrus and thus failed to reduce pest levels.[33] Despite this disappointment, *Habrolepis* did help to unravel the mystery surrounding the earlier introduction of *Comperiella bifasciata*.

Answers began to emerge shortly after Compere sailed for Capetown. Howard L. McKenzie, an assistant in the Division of Entomology at Riverside, finally found reliable morphological characters to distinguish among the species of *Aonidiella*, which he separated from *Chrysomphalus*. Compere had actually seen the microscopic sclerotized area that characterized *A. aurantii*, but he thought it also occurred in *A. citrina*. Museum specimens Compere used at the time were in fact all *aurantii*. Explorers had been dealing with not two but five species that often favored different plants within the same region. Because of California's quarantine against citrus imports (to avoid the citrus canker disease), explorers repeatedly shipped supposed red scale, parasitized by *Comperiella*, on an ornamental plant. McKenzie identified this scale as *Aonidiella taxus*. The parasite attacking it could switch to *citrina*, but not to *aurantii*. Even the *Chrysomphalus* species on which *Comperiella* had first been established in California turned out not to be the one entomologists thought it was at the time.[34]

The story did not end with McKenzie's classification of the scales. Sometimes red scale in the Orient really was red scale. Flanders tried rearing the new South African parasite *Habrolepis rouxi* on red scale infesting sago palms. He found that the parasite could not develop. The seemingly agreeable host plant on which Compere carried red scale to Asia in 1932 had been part of the problem all along. Sago palm enhanced the already substantial ability of *A. aurantii* to encapsulate and kill parasites. Any Asian parasite capable of developing in red scale on citrus would still have failed in Compere's palm cages. Flanders's finding reopened the question of host-specific races of *Comperiella*. McKenzie then examined Asian specimens of scale known to have harbored the parasite. He confirmed their identity as *A. aurantii*, true California red scale.

Although war prevented Smith from sending anyone back to China, a colleague living there shipped parasitized red scale on citrus to the USDA receiving station in Hoboken, New Jersey. Emerging parasites attacked scale on California citrus, which could return to California without being stopped by the citrus canker quarantine. As a result, Smith's team established a red scale-inhabiting race of *Comperiella bifasciata* in 1941, thirty-five years after the first attempt.

From then on, Compere and other protégés of Smith cited the episode to illustrate the need for more research in systematics to support biological control. Smith always urged each staff member to take up systematics of some parasitic group of insects. Unfortunately, *Comperiella* never did control red scale in California. Whatever value the parasite had could at least have been determined much earlier if taxonomy, including levels below species and characters other than structural traits, had not lagged.[35] Pests, too, were not always sufficiently sorted out. Until McKenzie cleared up the genus

Aonidiella, entomologists could not know whether a parasite attacked an important pest or not—although the parasites could tell the difference.

Even a complete catalogue of which morphological species attacked which would not have been enough to get the right form of *Comperiella* to California. L. O. Howard described the species, and indeed the genus, from a single specimen that supposedly emerged from California red scale.[36] Howard did not use—would not have thought to use—ecological or physiological traits in a description. Under typological, morphological systematics, the assumption would be that the species as a whole could or could not parasitize red scale. *Comperiella* could, if the original record of the "type" specimen were correct. Now, more than thirty years later, members of one species were found to differ in their host relations.

Such findings implied that a single dead specimen could not adequately characterize a species. This conclusion was a major tenet of the "new systematics" emerging during the 1930s. Biologists increasingly argued that a proper description would have to cover the variability of a species throughout its range. A writer must include not only morphological traits and (in the case of parasites) host relations, but also other ecological data and physiological and cytological characteristics.

These points were repeated throughout *The New Systematics*, an anthology edited by British biologist Julian Huxley and published in 1940. Advocates of the new approach realized that many existing descriptions not only failed as guides to applied work, but contained too little information to tell which of several forms the original writers had studied. Leading American entomologists hailed the book and urged their colleagues to heed its call. In the book itself and in the flurry of response, however, entomologists disagreed as to whether systematists would have to learn the new techniques or submit difficult cases to other biologists. All, of course, called for increased funding for systematics. Biological control workers welcomed allies in that effort.

Huxley pointed out that biological characters alone sometimes distinguished pests or effective biological control agents from

economically unimportant species. He and others admitted that the
sheer numbers of insects could delay the movement of entomological
systematics beyond its old ways.[37] Nevertheless, the practical mission
of biological control could drive basic research in systematics. Harry
Smith applauded Huxley for making that point with the examples of
Trichogramma and California red scale in his landmark book *Evolution: The Modern Synthesis* in 1942.[38]

The practical mission remained. Having sent *Habrolepis* to the
United States, where it would touch off the chain of discoveries
leading to the establishment of *Comperiella* on red scale, Harold
Compere spent the rest of his 1936–37 stay in Africa searching for
black scale parasites. Earlier explorers had been criticized for not
going out into the wild country and really exhausting the territory.
Compere enjoyed the road and the grand scenery, but remained a self-
described "sidewalk explorer." Groups such as the Lemon Men's Club
were eager to hear of dangerous African safaris. He would tell his
audiences, however, "Collecting beneficial insects is a work that can
best be accomplished if adventure is eliminated."[39]

Compere's real advantage over his predecessors was the
availability of air transportation. Planes cut shipping time to California
and improved the insects' chances of arriving alive. Also, black scale
had become more of a pest in South Africa, presumably because
introduced ants were driving parasites off to protect the honeydew-
producing scale insects. Local treatment of the ant problem allowed
large numbers of parasites to attack the suddenly unprotected scale,
making collection easy for someone looking at the right time.

Among parasites sent in years past without establishment in
California was *Metaphycus helvolus*. It had seemed unimportant in
controlling black scale in South Africa. Compere thought no better of
the parasite than earlier explorers had. In accordance with Smith's
policy, however, he sent anything that might help. Riverside was soon
abuzz over the effect of *M. helvolus* on black scale populations. Faster
to develop and yet longer lived than its once touted congener *M.
lounsburyi*, the new parasite appeared better able to bridge the gap
between generations of the scale. Black scale bounced back after

particularly cold winters, which killed off many of the parasites, but *M. helvolus* was still the best species to come in during the long battle against this pest. Because the parasite had turned out well after being so lightly regarded, California entomologists used it to argue for multiple introductions. The case showed that the single best natural enemy in a new environment could never be predicted ahead of time.[40]

Between his move into the university in 1923 and the cessation of foreign travel at the outbreak of World War II, Smith assigned an explorer to crops other than citrus only once. That was himself. Hawaiian and USDA colleagues sent him natural enemies of a few pests of other crops in the 1930s, but never with much success. As more assistants joined the staff, Smith busied himself with mathematical ecology—and his other hobby, an intense interest in college football. He became more detached from the day-to-day work with insects.

Smith had not done any foreign exploration for twenty-six years, but in 1939 he took the opportunity for a six-month trip to Europe. He sought enemies for pests of deciduous fruits grown in northern California. As in citrus, the initial targets were scale insects. The trip appealed to Smith in part because it offered the chance to spend a few days in England arguing population dynamics with Robin Thompson of the Commonwealth Institute of Biological Control. Smith also visited France, Austria, and Germany, then spent most of his time in Italy. The tension that would soon explode into war prevented him from traveling farther east as he had planned. He was fortunate to get out in time.

As for the insects, Smith shipped much material home, but nothing of real practical value. Like Compere before him, he could not resist one premature announcement of a valuable parasite, in this instance one that failed even to survive in California. Smith reported that some of the pests he had come to investigate were more serious in Europe than in America. Europeans, he wrote, should look for natural enemies west of the Atlantic. Smith decided that less exacting cosmetic standards in Europe were the only reasons entomologists there considered the insects to be under control. He did not suggest lowering standards in the United States. Thus ended the first serious attempt of the University of California to expand biological control beyond citrus.[41]

S mith also broadened the horizons of biological control by offering the only graduate training in the field. Graduate research, as well as visits by scientists from around the world, encouraged Smith's deepening interest in ecological and evolutionary topics. Students from the program comprised most of its next generation of scientists. Riverside had no courses as such except for short summer sessions. Undergraduates learned their entomology at Berkeley, although in the 1930s the university's new Los Angeles campus began offering introductory classes in the subject. Smith occasionally gave a lecture or two at Berkeley or UCLA. Graduate students also took most of their course work in the north but could choose thesis research with Smith or Quayle at the Citrus Experiment Station.

After Flanders, the first group of graduate students in biological control came in the mid-1930s.[42] Smith suggested thesis topics, but almost anything relating to natural enemies of insects was fair game. One of the first dissertations was a taxonomic study of the scale parasite genus *Coccophagus*. Others concerned mortality in laboratory populations, using Smith's ideas on density effects. If a topic pertained directly to a Citrus Experiment Station applied problem, the student could work on it full time. Otherwise he or she must assist the regular staff in the lab to earn the $50 monthly stipend. Some students were a little jealous of people who, like Flanders, received full-time staff salaries while doing graduate work. The university later abolished the practice.

Young entomologists faced poor prospects for getting jobs in biological control anywhere else during the Depression years. One early student, Laurence Jones, left in 1938 just short of his Ph.D. to take a USDA position that kept him in Riverside but away from biological control. He never returned to the field of his training. Donald W. Clancy, with Ph.D. in hand, also joined the USDA that year for work outside biological control, but in less than a year received a transfer to a parasite project. Clancy had grown up in a citrus orchard, graduated from Berkeley, and worked at the Orange County Insectary before completing his thesis. Smith directed him to study parasites of the beneficial green lacewings, a project Smith had wanted to follow up since his own USDA stint decades earlier. Clancy stayed in biological control throughout his USDA career and, in the late 1950s, even returned to citrus insects.

Some students' laboratory studies strayed from the station's usual narrow focus on citrus problems, because Smith desired to bring biological control up to speed in the increasingly theoretical world of population ecology. Janet Mabry measured the effects of crowding on reproduction in laboratory populations of species not associated with citrus, seeking confirmation of density-dependent action on population growth rates.

Mabry took her first college entomology course at UCLA from Alfred M. Boyce of the Riverside chemical control group. Boyce would lecture with hands still covered by the insecticide he had applied to his test plots early in the morning. Mabry was Smith's only female graduate student. She fit in well with the men, accompanying the outdoors-loving "Prof Harry" on backpacking trips. Gender, however, blocked her pursuit of a career in entomology. As historian Margaret W. Rossiter has shown, entomology was one of the most male-dominated agricultural science disciplines before World War II.[43] Shortly before Mabry finished her Ph.D., she married Boyce. The university's strict antinepotism rules then prohibited her from taking a job at the station, in biological control or anything else.

Another student who began his entomological study with Boyce was Paul DeBach. One day Harry Smith gave a lecture at UCLA, with illustrations such as a photograph of Harold Compere riding a camel in Africa. From that moment, DeBach recalled, he wanted a career in biological control. After finishing his bachelor's degree at Berkeley, he came to Riverside. As a starving graduate student in his first summer, he would sleep on a cot in the lab or out on the hill above the insectary.

DeBach collaborated with Smith on elaborate experiments, using housefly larvae and their parasites, to determine the effects of density on the percentage of hosts attacked. Smith and DeBach also devised a technique to verify the effect of natural enemies in pest control, excluding parasites from selected tree branches by means of organdy sleeves. This, the first of a series of exclusion methods DeBach would employ over the years, was Smith's practical response to criticism that biological control lacked scientific verification. Entomologists could not usually make tests of one plot with natural enemies and a control plot without them, because adult parasites could spread quickly into the control plots. After completing his doctorate, DeBach remained in Riverside only a short time as an assistant before taking federal government positions outside biological control. He gladly took a salary cut to rejoin Smith as a regular staff member in 1945.[44]

Smith's students found few opportunities to continue careers in biological control anywhere but in his own program. The USDA could take people now and then as in Clancy's case, but could not promise to keep them in parasite work. State experiment stations other than California's lacked any ongoing programs in biological control. Even the University of Hawaii was only a minor player compared to the sugar planters, until after the war. Still, entomologists across the country recognized Smith as a leader, and elected him president of the American Association of Economic Entomologists for 1940.

It is difficult to say just how Smith rose to the top of his profession when the rank and file had little use for what he was doing. His efforts in population theory set him apart from most economic entomologists in the United States, but he maintained a practical outlook that other members of the AAEE must have appreciated. They knew that Smith was truly one of them and not some snobbish intellectual who might belittle the scientific standing of pest control. The work of others in his division, particularly Stanley Flanders, may also have enhanced Smith's reputation. Former associates described

Smith as such a kindly person that it would be difficult to imagine him having an enemy in the profession. However, in his presidential address in Philadelphia that December, he told the assembled insect killers things most were not yet ready to hear.

Association presidents usually gave comfortable presentations summarizing their own fields or triumphantly reflecting on the accomplishments and importance of economic entomology. Drawing on his experience in California citrus and on his enthusiasm over developments in evolutionary theory, Smith told his colleagues that insects would not just sit and be conquered, but instead might change to avoid any weapon directed at them. His address, "Racial Segregation in Insect Populations and its Significance in Applied Entomology," though not about biological control, dealt with issues raised in the course of his work and warned against exclusive reliance on chemicals.

Smith's definition of "race" was little more precise than the one his old postdoctoral fellow William Thorpe had used in 1930. Thorpe had included what were later called "sibling species." By 1940, as part of the developing evolutionary synthesis, biologists increasingly recognized the category of species as something qualitatively different from other taxonomic levels. Speciation was that point in evolution when two groups descended from one species could no longer interbreed and produce fertile offspring. In genetic terms, this meant that genes could not be transferred between the two forms. Biologists would now restrict usage of the term "species" to reproductively isolated forms. The size and number of genera and other taxonomic levels above the species remained arbitrary, for they lacked any neatly definable, qualitative criterion for separation. "Race" would apply only to forms within a species, differing in some respects but still capable of interbreeding. Thorpe made his own conversion to this usage clear in his contribution to *The New Systematics*.[45]

The individual most responsible for bringing biologists to this consensus was the Russian-born, American geneticist Theodosius Dobzhansky. Dobzhansky possessed a field biologist's curiosity about the variability of natural populations, a systematist's concern for telling species apart, and a solid background in genetics. He focused on the evolutionary question of how species come to be. Others, notably British geneticist William Bateson, had stressed the importance of reproductive isolation in defining species. Dobzhansky, however, emphasized species as a stage in the evolutionary process, precisely definable yet sometimes impossible to demarcate in practice because complete isolation would require many generations. Incompletely isolated intermediates between race and species necessarily complicated the systematist's task. Especially in his book *Genetics and the Origin of Species*, published in 1937, Dobzhansky pinpointed the question at the center of evolutionary biology: not how species differed, but how they *became* different.[46]

The success of Dobzhansky's book lay in its appeal to various biologists who shared one or another of his diverse interests. The main attraction for Harry Smith was Dobzhansky's discussion of the process by which a species divided into races.

It is not clear whether Smith ever meant sibling species when he said "race." By 1940, in any event, he did not. He went to the other extreme and included single genetic differences such as the factor for cyanide resistance in the California red scale. In his view, then, races of red, black, and citricola scales had formed and spread during his own career in California. Entomologists at Riverside demonstrated a Mendelian genetic basis for the change in red scale. These examples were familiar, if not unanimously accepted among pest control workers.[47]

What spurred Smith to emphasize chemical resistance and other subspecific variation in 1940 was his reading of *Genetics and the Origin of Species*. Dobzhansky used the extensive genetic variation he had found in natural populations of insects to draw implications for the evolution of new species. He also called the spread of resistance to fumigation in the three scale insects "probably the best proof of the effectiveness of natural selection yet obtained." At one point Dobzhansky used the word "race" to describe resistant scale.[48] Smith, in his presidential address, highlighted the formation of races as a potentially rapid prelude to the slower process of speciation. The latter awaited the evolution of what Dobzhansky had termed "isolating mechanisms," the physiological means by which populations were prevented from interbreeding—and the very heart of his book.[49]

How well Smith and Dobzhansky knew each other in 1940 is unknown. Dobzhansky's association with Riverside began almost immediately after he moved from Columbia University to the California Institute of Technology with Thomas Hunt Morgan in 1928. The Citrus Experiment Station had the largest concentration of professional entomologists in southern California. An entomologist by training, Dobzhansky used collections at the station. He may also have passed through Riverside often on the way to *Drosophila* collecting sites in the mountains to the east. Quayle and his assistants consulted Morgan and Dobzhansky about cyanide resistance shortly after the geneticists arrived in Pasadena.[50] In his book, Dobzhansky cited Quayle's unpublished data.

Evolutionary topics were rarely of prime concern to agricultural entomologists going about their usual studies of how to kill insects. The experience of cyanide resistance among citrus insects, the taxonomic difficulties and possible races in scale insects and their parasites, and a personal acquaintance with Dobzhansky and Thorpe would all help direct Smith to the subject of evolutionary change. His work in population dynamics already reflected a desire to be where the action was in biological science. While it was unusual, it was not unheard of for an economic entomologist to explore evolutionary mechanisms. A few, beginning with biological control zealot Benjamin D. Walsh in the 1860s, had considered the case of host-specific phytophagous species.

Walsh called host-specific races "phytophagic varieties," and host-specific sibling species "phytophagic species." If a species strongly favoring a particular plant happened to lay eggs on another, he argued, any surviving offspring would be more likely to breed on the new host. Walsh did not specify whether this would occur by natural selection—survivors being those best fit for the new host—or by acquired preferences becoming hereditary. The latter, Lamarckian mechanism seems to be what he had in mind. If many generations lived on the new host plant without interbreeding with individuals from the original host, a variety preferring the new plant would become a species requiring it. Lacking the time to watch this happen, Walsh argued from a series of cases that, he felt, represented a continuum of intermediate steps in sympatric speciation. In general, he wrote, other differences would evolve to make reproductive isolation permanent even before host preferences became obligate.[51]

Unaware of Walsh's discussion of the subject, Frank C. Craighead of the Division of Forest Insects in the U.S. Bureau of Entomology made the same argument in 1921, also implying inheritance of acquired characteristics but not stating it explicitly. Craighead made a series of experiments forcing insects to change plant hosts, in order to verify a statement of his division head, Andrew D. Hopkins. Hopkins simply suggested that, in a species with distinct varieties living on different plants, individuals would be adapted to the host on which their ancestors lived, and would tend to continue breeding on it. Hopkins said nothing about how this adaptation came about. Craighead named the phenomenon the "Hopkins host-selection principle," and did discuss the origin of such varieties. He noticed that attempts to force host-specific insects to switch plants usually caused high larval mortality. He did not suggest that the improved ability of survivors to live on the new plant resulted from selection of those already most suited to it. Craighead's explanation, like Walsh's, seems Lamarckian, although one cannot be certain.[52]

Writing in 1930 after completing two years in Smith's lab, William Thorpe reviewed the subject of biological races. He included non-interbreeding but morphologically indistinguishable forms. Thorpe thus rejected the opinion of Walsh that systematists should consider these true species—the view that would prevail after another decade. Thorpe's examples included California red scale. He believed that it had an Asian race, susceptible to the parasite *Comperiella bifasciata*, and an immune California race. He argued that Craighead's races need not be genetically different. Instead, developmental effects of the plant on the larva could cause it to return to the same plant to breed as an adult. Thorpe ascribed this "larval memory" effect to Walsh, although

Walsh's statements are not clear on the issue. Larval memory could still serve evolution by keeping populations apart long enough for speciation to occur. Thorpe was not ready to decide on the question of Lamarckian inheritance. He doubted that selection explained chemical resistance, because resistant red scale had not spread to all fumigated areas.[53]

Smith expanded the discussion to include races of parasites specific for their insect hosts. The evolution of such forms bore directly on biological control. Smith suspected by 1940 that, instead of having differential success on races of the host, *Comperiella* existed in one form that could attack California red scale, and in one that could not. This would soon prove to be the case. Their ability to interbreed confirmed that the two forms were merely races and not sibling species. Flanders's work with *Trichogramma* seemed to involve host-specific parasite races. In another familiar California example, a new plant-specific race of codling moth formed when the moth switched from apples and pears to walnuts. That was why Flanders started studying *Trichogramma* in the first place.[54]

Exactly when Smith accepted natural selection to the exclusion of acquired characteristics is not clear. Early in his career, he mildly echoed the old California enthusiasts' dictum that newly introduced species would become more effective after they had a few years to adapt to the new environment. This argument, like those of Walsh and Craighead, seemed implicitly Lamarckian. It lacked any reference to natural selection, but did not address the mechanism for the supposed change.

Such statements disappeared, without comment, from Smith's writings long before he took up insect races. He never knew an introduced natural enemy to become more effective years after its establishment. By 1937 he expressly rejected claims of such instances. After reading Dobzhansky, if not before, Smith flatly rejected the apparently Lamarckian interpretations of Craighead, or even Thorpe's "larval memory." Smith found it unnecessary to resort to such an explanation. Indeed, the ebbing of Lamarckism seems to have occurred largely through such realizations. Inheritance of acquired traits was simply not needed to explain supposed examples. Thorpe distanced himself from Lamarckism in the same manner following his own studies of sensory conditioning in the late 1930s, although he was reluctant to dismiss the possibility altogether.[55]

In addition to Lamarckism, saltationist views of speciation receded from orthodox theory as the evolutionary synthesis joined genetics to natural selection. With the demise of saltationism, and with Dobzhansky's new emphasis on physiological isolating mechanisms, came the restoration of a pre-Mendelian consensus that geographic isolation must precede the formation of distinct races and species. Smith argued that it would take some distance between populations, for example between walnut and apple orchards, to prevent interbreeding long enough for a new host to select characteristics fit for the new conditions. Even more time would be required to develop biological isolating mechanisms that could keep the populations distinct after the spatial gap was removed. Smith thus doubted that host preference provided a truly sympatric process of speciation, although Thorpe and even Dobzhansky held out the possibility that ecological isolation might be accomplished without geographic separation. Reviewing the subject a short time later, Ernst Mayr cited a distinct shortage of evidence for this type of sympatric speciation.[56]

Natural selection required variation. Others in biological control, particularly Smith's former assistant and future successor Curtis Clausen of the USDA, had begun to see the practical implications of genetic variability within a species. Clausen recommended collecting a natural enemy throughout its geographic range lest an unknown race, better suited than others for the new environment, be neglected. Dobzhansky's findings of variability and of short-term genetic change in natural and laboratory populations pushed Smith a step further. He cited Dobzhansky repeatedly in the 1940 address to the American Association of Economic Entomologists.

The most desirable race need not exist already, Smith argued. The genetic diversity of any large population contained the potential for new races to be separated out by natural selection in any new or altered environment. Just this had occurred when frequent fumigation changed the environment of red scale, and after codling moth happened to spread into walnut orchards. If the original, susceptible form of scale still survived, said Smith, it must be more fit in the absence of fumigation. Susceptible scale would increase in frequency between fumigations and continue to dominate areas with little or no use of the gas.[57]

Given Smith's belief in potential races and his newfound recognition of the work of geneticists, he might have called upon economic entomologists to learn genetic techniques in order to identify Mendelian factors affecting the value of biological control agents. Instead, he merely advised his colleagues to notice what geneticists had done. Others, however, would attempt to improve parasite populations by selective breeding.[58]

Perhaps Smith, then nearing sixty, could not consider such a disciplinary upheaval. Like others in his profession, he was more of a field naturalist than a laboratory biologist. His earlier inattention to genetics is understandable. Laboratory geneticists tended to study clearly visible traits in pure lines of a small number of species, with little regard to their economic importance or even adaptive significance. Geneticists often excluded talk of natural populations.[59] Smith realized that the visible characteristics early geneticists had favored were much easier to analyze than the physiological and behavioral traits affecting biological control. "From the standpoint of applied science," he said, "they are far more important, since it is largely they and not the morphological and color variations that determine what an organism *does*."[60]

Faithful to his mission of helping orange growers, Smith would have had little use for a discipline that neglected wild populations of parasites and predators, until Dobzhansky forced his attention to genetics. Influence flowed in both directions. Dobzhansky, one geneticist who did work extensively with natural populations, expanded his discussion of pesticide resistance when revising his book. Dobzhansky cited Smith's recent address and personal communication with Smith in the new edition, which appeared in 1941.[61]

When Smith did consider the practical implications of evolution and genetics, he did not return to the old supposition that a new natural enemy ought to improve after a few years. His concern now was the ability of pest populations to respond to manmade changes in the environment, especially the use of pesticides. "There is every reason to believe," he warned his assembled colleagues, "that this condition will develop with many other pests, and the economic entomologist can expect to be continually faced with the prospect that many of his recommendations will, in time, become of doubtful value." As he had elsewhere urged entomologists to pay more attention to mathematical ecology, Smith now urged them to act on the implications of genetics and evolutionary theory. Pest control specialists must learn the biological mechanisms of resistance, in order to devise treatments to get around the problem. Some treatments would have to be restricted if they were to retain any effectiveness.[62]

Of course Smith's own specialty, biological control, would have to receive more attention if overuse of chemicals were headed for the fall he foresaw. Examples of resistance to chemicals might have been relatively few, but *no* insects were known to have newly evolved resistance to a natural enemy. In principle, it could happen in the very long run—witness the ability of red scale to encapsulate developing parasitic eggs—but presumably this would be far less common than pesticide resistance. Besides, natural enemies could also evolve: A race of *Comperiella* overcame the red scale's defenses. Smith had never claimed that biological control could solve all agricultural pest problems, and he was not going to start now. He had always argued that native pests were poor candidates for biological control. In his presidential address, he reaffirmed that growers must not cut back their pest control treatments. He challenged entomologists to improve those treatments.[63]

Smith recognized that his role in agriculture exactly matched that of his chemically oriented colleagues who occupied the floor below his in the entomology building at Riverside. They all aimed to make a spotless orange and maximize the growers' profits. Smith explicitly acknowledged this purpose earlier in his career; his growing awareness of chemical resistance did not change his views. He dismissed contentions that pest control measures should be curtailed in times of agricultural surplus such as during the Depression. In 1941 he wrote that growers could not be blamed for taking drastic steps with insecticides when a pest seemed to be getting out of hand, even if the treatment killed parasites that might have solved the problem. Smith gave the standard economic entomologist's position: Any crop loss to insects was a waste, regardless of surpluses.[64]

Addressing the nation's economic entomologists out of his new interest in evolution, Harry Smith anticipated one of the detrimental effects of insecticides. In the next few years, during and shortly after World War II, entomologists eagerly adopted a new generation of chemicals that would bear out Smith's gloomy predictions of overuse and resistance. But he did not abandon his outlook. Profits for the growers took precedence. He did not publicly express any concern over environmental damage or danger to consumers from pesticides. He admitted in 1946 that relaxed cosmetic standards could reduce the use of chemicals and help forestall the evolution of resistance. In the same breath, however, he dismissed the idea. The high capitalization and competitive nature of the citrus industry would not permit it.[65]

Most economic entomologists were not ready for Smith's call for restraint. Even as he issued that call, experiments were beginning to uncover the remarkable insecticidal properties of a chemical to be known as DDT. Smith pointed to exciting developments in biology, namely ecology and evolution, but his profession turned more toward chemistry. On the eve of the war and of the arrival of new pesticides, Smith's organization still focused on the insects affecting a single crop. The scope of scientific activity within the Division of Beneficial Insect Investigations had broadened. During and after the war, even as DDT came on the scene, the practical agenda of Smith's group grew as well. The division finally began to expand beyond citrus in a permanent way, including the establishment of a branch at Berkeley to handle northern California problems. Besides moving into more crops, Smith added weed control and the use of microorganisms as control agents.

The chemical onslaught, however, retarded any interest other states might have had in joining a biological control movement. And in the U.S. Department of Agriculture, biological control all but collapsed just when it was needed most. California would still have to go its own way until the insects forced others to follow. When that time came, internal strife crippled California's ability to lead.

Vedalia beetles feeding on cottony-cushion scale, c. 1950. *Photo by K. L. Middleham.*

Photographs from the files of the Department of Entomology, University of California, Riverside, courtesy of T. W. Fisher, Specialist Emeritus.

French entomologist Paul Marchal and U.S. Bureau of Entomology Chief Leland O. Howard (*right*) in Salt Lake City during Marchal's tour of American biological control facilities, 1912.

Harold Compere rides a camel while exploring for black scale parasites in Eritrea. Paul DeBach credited this photograph with inspiring his own choice of a career in biological control.

Black scale, a major target of California biological control efforts for decades, nearly covers this branch.

Stanley E. Flanders searches for citricola scale parasites among wild lime trees in Australia, 1931.

Harry S. Smith examines sago palms in the insectary at Riverside during Harold Compere's 1936–37 trip to Africa. Compere used sago palm to transport red scale on several of his voyages, but this host plant later proved unsuitable for the parasites.

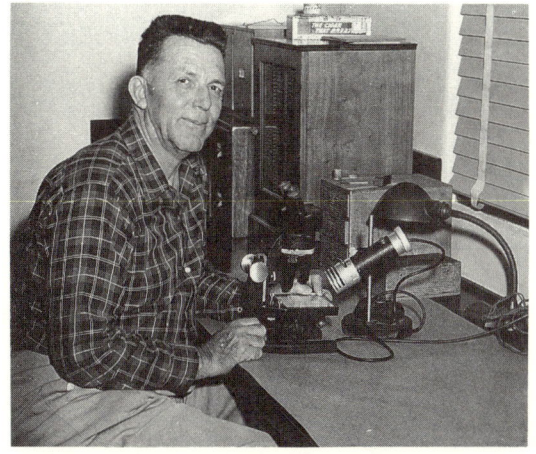

Stanley E. Flanders, long-time insectary supervisor and student of parasite development, at his desk in Riverside, 1955. Flanders profited greatly from his own citrus orchards near the campus—by selling them off for subdivision. *Photo by Francis A. Gunther.*

Philip H. Timberlake in his most familiar position, at the microscope and surrounded by insect boxes, 1955. Already five years past retirement, he would remain at work for another quarter century. *Photo by Francis A. Gunther.*

Curtis P. Clausen, second chairman of the University of California Department of Biological Control (1951–58), attends to administrative duties. He was said to have brought a more bureaucratic approach from his long service in the federal government. *Photo by Francis A. Gunther.*

Agricultural science often meant getting one's hands dirty, as it did for Blair R. Bartlett in this 1955 photograph. Bartlett applied the phrase "integrated control" to the combination of chemical and biological approaches. *Photo by Francis A. Gunther.*

Charles A. Fleschner at the Citrus Experiment Station in 1955. These were the happier days before his tumultuous tenure as third and last chairman of the statewide Department of Biological Control (1958–63).

For much of his career, Irvin M. Hall was the only insect pathologist at Riverside. When the statewide department was broken up, Hall chose to stay in the Riverside campus Department of Biological Control, of which he was the last chairman (1966–69). *Photo by Francis A. Gunther.*

William Robin Thompson, director of the Commonwealth Institute of Biological Control, on a visit to Riverside in 1949. An early colleague of Harry Smith, Thompson tried to mathematize population dynamics, but he later became a vehement critic of such efforts.

Paperwork continued to burden Harry Smith after his retirement, as seen in this 1955 photograph. He never completed his projected text on biological control. *Photo by Francis A. Gunther.*

Members of the Department of Biological Control pose outside Harry Smith's desert cabin during the 1948 statewide meeting. *Back row, from left:* Carl B. Huffaker, Kenneth M. Hughes (*kneeling*), Theodore W. Fisher, Almon J. Basinger. *Middle row:* Stanley E. Flanders, Glenn L. Finney, Blair R. Bartlett, Kenneth S. Hagen, Harry S. Smith, Wendell F. Sellers, Harold Compere, Edward A. Steinhaus. *Front row:* Charles A. Fleschner, Philip H. Timberlake, Everett J. Dietrick, and Paul DeBach. *Photo by James K. Holloway, courtesy of the Department of Environmental Science, Policy, and Management, University of California, Berkeley, and the Bio-Integral Resource Center.*

CHAPTER 6

Growth and Conflict in
the Age of DDT

The years immediately following World War II were the golden age of chemical pesticides. A barrage of new compounds, led by the spectacular DDT, swept across American farms with the promise of ending insect problems completely and permanently. The most enthusiastic entomologists talked of driving even widespread pests extinct. A few warned of the danger ahead, but their voices nearly drowned in the euphoria over miracle chemicals. The message went out that farmers need not tolerate insects in their fields, and consumers need not tolerate spots on their oranges or holes in their lettuce.

State experiment stations outside California had largely ignored biological control even before DDT, parathion, and the new breed of chemicals came along. The U.S. Department of Agriculture scattered its corps of biological control workers among divisions that concentrated on pesticides. The war handcuffed beneficial insect work; enthusiasm for chemicals threatened to eliminate it entirely. Biological control took too much planning and supervision, and had a low success rate.

Exaltation of pesticides soon had to be tempered. DDT did not fulfill its early promises. It killed insects, to be sure, but not all insects, and not always the right ones. The bugs fought back, evolving resistance to DDT in as little as two years and then keeping in step with almost anything the chemical industry threw at them. Economic entomologists responded to each round of resistance with another poison, which became ineffective in turn. Insects previously under complete biological control—and some that had never been known as pests—broke out in menacing numbers following the use of broad-spectrum insecticides. The usual answer was to add a second chemical, rather than retreat from the first one.

Public fears of danger to human and animal life flared sporadi-cally. Rachel Carson published *Silent Spring* in 1962 and triggered a national pesticide hysteria, which culminated ten years later in a federal ban on DDT. By 1962 the USDA and others had already begun to reconsider their commitment to the all-out, chemical approach to pest control, but the profession of economic entomology condemned Carson almost unanimously.[1]

California was at the forefront of the chemical explosion after World War II. The nation's most diverse and productive agricultural state liberally supported insecticide research and greedily consumed the resulting potions. Grants from the chemical industry helped, but the state funded most of the rapid growth of entomology divisions at Berkeley and Riverside. California accelerated other forms of modernization, such as mechanized harvesting. The agricultural science complex came under fire in the post-Carson era for disregard-ing small farmers, farm workers, and the public interest. Critics charged that such developments gave large-scale, corporate farming an undue advantage over smaller operations and threatened the quality of the food supply while needlessly increasing its quantity in an age of rampant surplus.[2]

Biological control survived the onslaught of chemicals. Although at first the new compounds seemed to reduce the importance of beneficial insects, pesticides were no more successful at eradicating the University of California biological control team than at wiping out the boll weevil or the codling moth. While the eclipse of biological control in other states and in the USDA became more complete, California's beneficial insect program joined the postwar expansion of government-sponsored science. Biological control grew from one team concentrat-ing heavily on citrus at an out-of-the-way substation to a two-campus department with dozens of Ph.D. scientists.

Yet, when the popular backlash against insecticides arose in the 1960s, most members of California's Department of Biological Control remained as silent as the spring of Rachel Carson's hyperbole. They were economic entomologists, not environmentalists. Although Carson praised the methods they happened to study, the "parasite men" in Riverside and Berkeley were as wedded to modern, industrial agriculture as were their counterparts in chemical control. Biological control workers merely believed that in certain cases their way could also produce spotless fruit—but at less cost in the long run. Each camp aimed at the same goals: agricultural profit and productivity.

Scientists also insisted that it was not their fault if their very success at coaxing more and prettier produce from the land actually drove farmers out of business. That entomologists should be concerned with broader issues seemed preposterous to most, until public pressure forced a change in perspective. Scientists did not appreciate being attacked for simply doing their job.

California retained an active program in biological control partly because of a few conspicuous, past successes and a steady campaign of reminders. Cottony-cushion scale once threatened to destroy the citrus industry, until imported natural enemies saved the day in 1889. The memory of that incident sustained biological control through long years without similar accomplishments. Control of citrophilus mealybug forty years later, along with a handful of partial victories, kept the faith alive. Citrus growers, beneficiaries of most of this work, also were among the first to see pest insects become resistant to chemical treatments. The organization of agriculture and of agricultural science may have been more amenable to biological control in California than elsewhere. But the same factors contributed to the state's leadership in chemical control, and to an increasing awareness that the two methods must be harmonized to some degree. The latter movement became known as "integrated control."

The continued existence of a biological control program, and its expansion to new crops and new methods, reflected the commitment of its leaders to modern, organized, industrial agriculture. Because of that commitment, California's biological control scientists generally spurned an environmentalist movement that would have welcomed them. They did not condemn insecticides for poisoning wildlife or endangering consumers, but asked only that natural enemies be allowed a chance at susceptible pests, without interference from chemicals used irresponsibly.

The program grew almost entirely from within. As it opened new scientific jobs, its own students filled them. Institutional inbreeding strained what had once been almost a family atmosphere. Scientists grew jealous over authority and territory. A biological control colony established at Berkeley in the 1940s yearned for independence from Riverside in the 1950s. Finally, in the 1960s, the Department of Biological Control blew itself apart, not because of competition from pesticides but because of a storm of internal dissension, which university administrators failed to calm. The team approach that had been the small program's strength gave way, perhaps inevitably, to the desires of too many individuals who had been together too long. A period that opened with expansion, in the midst of World War II, closed in virtual disintegration and the loss of institutional identity.

The department collapsed at what should have been a time of unprecedented opportunity for biological control. Through the integrated control trend, greater numbers of entomologists gave at least lip service to the importance of natural enemies and admitted that chemicals sometimes did more harm than good. The "ecology" movement, which would later settle on the term "environmentalism," was gathering strength. Sparked especially by *Silent Spring,* agitation over pesticides loomed large in that movement, yet most scientists who worked in biological control steered clear. "Biological control" came to mean many things it had not meant through most of its history. Pest-resistant crops, sterile insect releases, or even old-fashioned cultural controls—just about anything other than chemical pesticides—came under the political banner of biological control.

Such methods were outside the scope of the Department of Biological Control of the University of California. Pest control had a new public, one that entomologists did not want. They clung to the farmers as their only constituency. Entomologists' reluctant and still incomplete acceptance of responsibility for people, places, and events beyond the farm is a story for others to tell. The immediate postwar history of the Department of Biological Control completes my discussion of an era in which agricultural scientists could confidently ignore those issues.

W orld War II exerted its most immediate effect on biological control by halting foreign exploration. Harold Compere, who had traveled the world in search of natural enemies of citrus pests since 1927, had to stay at home and find other things to do. Without new material coming in for testing, the rest of Harry S. Smith's Division of Beneficial Insect Investigations also slowed.

Smith wanted each staff member to take up systematics of some parasitic group as part of his regular work. Compere had done so between trips. The grounded explorer now turned full time to the taxonomy and morphology of parasitic Hymenoptera. Still smarting from his earlier failure to spot the characters separating red and yellow scale and to identify certain parasites on the road, Compere honed his skills during the break from travel. He hoped to clean up some of the errors that had wasted his and others' time abroad and in California orchards. Especially after his exploration career ended permanently in 1951, he preached the importance of sound systematics as the basis of biological control.[3]

During the wartime lull in exploration, Compere also purchased an orange grove in Riverside and became a businessman-farmer. He entered the business just as the shackles of the Depression came off. Complaints of surpluses were temporarily silenced. All-out production again became the goal of the citrus industry. Growers were warned that, as always, success of the individual and of the whole industry depended on shipment of only top-quality—that is, cosmetically perfect—fruit. Despite improvement in the market, no slackening in pest control could be tolerated. Compere adhered to strict cosmetic standards, because otherwise he could not sell his crop at a profit. Admitting that scarred oranges were just as delicious as shiny, spotless ones, he nevertheless scorned inattention to appearance in countries he had visited. Finding himself "successful as a grower of Red Scale and Red Spider," he readily used insecticides. Compere's attitude represented no change. The biological control team at Riverside had always agreed to high cosmetic standards.[4]

Although its industrial outlook remained in place, the program at last broadened its activities beyond citrus and beyond the Riverside campus. Oriental fruit moth (*Grapholitha molesta*), already a major pest of stone fruits in the eastern United States, was discovered in California in 1942. The insect threatened northern crops, although it first appeared in the southern part of the state. Legislators quickly made a large emergency appropriation to meet the threat.

The project was more service than research. No new parasite introductions were contemplated. The USDA had tried them for ten years and failed completely. Federal entomologists now relied on mass production of a native American parasite, *Macrocentrus ancylivorus*. To build up *Macrocentrus* production in California, Smith sent several assistants to the university's Gill Tract in Albany, three miles from the Berkeley campus. They soon developed a technique far cheaper and more efficient than that used in the eastern states, with a different host insect. The USDA adopted the California process immediately.[5]

As was often the case in biological control, research with a specific economic objective uncovered fundamental scientific information. Winter diapause, or hibernation, of host moth larvae prevented year-round production of *Macrocentrus*. Breaking the diapause required knowing what caused it in the first place. An assistant determined that length of night, rather than length of day as in cases studied previously, triggered hormonal changes leading to diapause. The practical result was that manipulation of photoperiod by artificial light or blocked windows could prevent the changes.[6] Instead of waiting for basic science that might someday find application, agricultural need directed fundamental research. The commitment to agriculture was neither a burden to Smith and his staff, nor merely an excuse for doing science, but was a guiding purpose that led to both service and research.

Oriental fruit moth never did become the scourge expected in

California. The Division of Beneficial Insect Investigations discontinued *Macrocentrus* production, but established a permanent foothold at Berkeley. For several years, university administrators had wanted to expand biological control beyond citrus. USDA explorers had sent natural enemies of pests of other crops to California on occasion. Little came of these efforts. Indeed, aside from citrus, introduced natural enemies received approximately the same inattention in California as in other states. Smith built his program around citrus. Close contact with growers and thorough follow-up of introductions in the field were lacking in other crops. The small staff focused on one agricultural sector eager for the service rather than spreading itself thinly.

Claude B. Hutchison, dean of the College of Agriculture and a friend of Smith, especially hoped to change this narrow perspective. Hutchison had encouraged Smith's trip to Europe in 1939 in order to bring northern California crops into the program.[7] The war set the effort back by cutting off exploration, but the oriental fruit moth emergency brought new funds and facilities into the division.

Less than a mile from the shore of San Francisco Bay, the Gill Tract provided a better environment and more convenient location than semidesert Riverside for work with insects of northern California crops. The colony at Berkeley-Albany made the Division of Beneficial Insect Investigations a truly statewide entity. Its headquarters remained at the Citrus Experiment Station. Smith directed operations at both sites. After the initial move for *Macrocentrus,* it was not long before the northern staff matched Riverside's in number and exceeded it in topical breadth. Permanent expansion began with two phases quite different from the division's traditional work on entomophagous insects in citrus. Before the end of the war, Smith added biological control of weeds and a semiautonomous unit for insect pathology. Both went to Berkeley.

The study of natural enemies was already more welcome there than at most agricultural campuses around the country. American economic entomology at midcentury has been characterized as more applied chemistry than applied biology.[8] This assessment echoes early California officials who derided "kerosene entomologists." At

Berkeley, however, entomologists had always cautioned against the overuse of chemicals.

In particular, Abraham E. Michelbacher was already practicing what would be called "integrated control," which many considered new in the 1950s. He helped Smith's team introduce parasites of the alfalfa weevil to northern California in the early 1930s. To preserve natural enemies and to control pests with minimal intrusion, Michelbacher called for crop rotation and judicious use of chemicals. Although the Berkeley group did not formally engage in biological control, which remained the province of Riverside until the 1940s, the university's flagship campus was not hostile to the idea when Harry Smith moved part of his team north.[9]

Smith was a conservative, frugal person, both in his personal life and in his university service. He did not want to gamble the taxpayers' money on ventures he believed had little chance of success, and he said so. Alfred M. Boyce, his counterpart in the Division of Entomology at Riverside, recalled that Smith's frugality held biological control back. According to Boyce, Smith had to be talked into new departures such as insect pathology, weed control, and even *Macrocentrus* mass production in the 1940s.[10] Boyce had the perspective of a dynamic leader and very successful fundraiser who got everything he wanted in the postwar period. Almost anyone could look timid in comparison to Boyce, but his claim that Smith resisted these new projects does not withstand scrutiny.

Having spent years playing down exaggerations of biological control, Harry Smith could not ask for money he did not believe would be well spent. Since he did not play budgetary games, he often had money left over at the end of the year. Spending money one did not have was anathema to Smith, who opposed the New Deal for that reason. When he had confidence in a project, however, he would not hesitate to lobby personally in Berkeley or Sacramento. He defended new facilities at Albany for the oriental fruit moth as a gamble worth taking.[11] When an opportunity arose largely by accident, he added insect pathology. Smith advocated the weed project for more than a decade before federal authorities finally allowed him to give it a try. That he did not match Boyce in pushing the limits of expansion was due more to Boyce's exceptional strengths than to Smith's weaknesses.

Smith took on insect pathology after a long period of doubt that artificial use of microorganisms could aid in controlling insects. Others, including his mentor and father-in-law, Lawrence Bruner, had attempted to spread fungi artificially since the latter part of the nineteenth century, especially against grain pests in the Midwest and citrus insects in Florida. These efforts never achieved any practical success. Smith especially doubted that fungi could help in California's relatively dry climate.

Smith's perspective on microorganisms changed after USDA agents discovered the milky disease (the bacterium *Bacillus popilliae*) of the Japanese beetle. Nearly twenty years of importations of parasitic insects failed to control the beetle in the eastern states before the USDA came upon the bacterium and found that it could be cultured and spread artificially. The discovery reawakened interest in what came to be called "microbial control."[12] Smith looked for a way to bring it into his division. He got more than he bargained for.

Neither Smith nor anyone on his staff knew much about insect microbiology. In 1944, need dovetailed with opportunity. Edward A. Steinhaus, trained at Ohio State University in both entomology and bacteriology, came to Berkeley to work on insect-borne diseases such as plague and encephalitis. Within a few weeks, Smith asked him to look at a disease that interfered with parasite rearing for oriental fruit moth at Albany. Steinhaus helped solve the problem. By the end of the year, Smith was recruiting him to develop microbial control. Smith's friend Hutchison helped move Steinhaus from the bacteriology department to head a Laboratory of Insect Pathology within the Division of Beneficial Insect Investigations. Steinhaus borrowed space from the Berkeley entomology division and conducted most of his operation on the main campus, rather than at the Gill Tract.[13]

Steinhaus came in on Smith's initiative, not because somebody else pushed for microbial control. Steinhaus's presence at Berkeley gave Smith the chance to start a program he had wanted for some time.[14] Once Steinhaus joined the division, however, he continually asked for more space and more money for insect pathology. He insisted that any practical benefits must await a good deal of basic

research, and that basic research was an appropriate end in itself. He
wanted to study not just microbial diseases of pests, but all ways in
which things went wrong in an insect or any other invertebrate.[15] The
Berkeley campus increasingly emphasized fundamental over applied
research. This environment suited Steinhaus better than Riverside,
where science was still seen very much as a means to an agricultural
end. Although he supported the application of microbial control,
Steinhaus's ambition went far beyond the agricultural commitment of
Smith.

Despite the disclaimers, the first practical results were not long
in coming. In conjunction with Berkeley entomologist Michelbacher's
alfalfa studies, Steinhaus and his students found that a virus disease of
the alfalfa butterfly (*Colias eurytheme*) could be cultured and spread
artificially for control. Cumbersome culture methods and a long
incubation period limited the value of the virus. Steinhaus tried
bacteria and discovered, to his surprise, that an old laboratory culture
of *Bacillus thuringiensis* killed caterpillars overnight. It proved
effective against many other insects as well.

Steinhaus championed the use of "*B.t.*" as a biological insecticide, although it had not required much of the basic research he
cherished. He pleaded with chemical companies to commercialize a
product that would be applied like chemicals but without harmful side-
effects. The industry finally came around after he made his case in an
agricultural chemical journal in 1956. *B.t.* went on to widespread use.

Few other microbial products followed, however. Steinhaus's
group did very little exploration for new diseases. The Laboratory of
Insect Pathology stood in sharp contrast to the rest of the division,
which Smith renamed the Division of Biological Control in recognition
of its spread beyond beneficial *insects*. Steinhaus objected that the new
name still stressed application rather than basic research, but for Smith
this choice was obvious.[16]

The biological approach to weed control more closely resembled
what the rest of the division was doing. California led the way in
control of insects by other insects and, despite the late start, led also
in microbial control. On weeds, however, California moved only after
practical achievement occurred elsewhere. As in the work on citrus

insects for so many years, calls from an agricultural industry spurred biological control of unwanted plants.

Livestock owners suffered losses from range weeds and pressed the University of California to import weed-eating insects, as had already been done in Hawaii and Australia. Hawaiian entomologists, including biological control pioneer Albert Koebele, introduced insects to control the exotic range weed *Lantana camara* early in the century. Touted as a substantial success, the effort later proved to be of little practical value, requiring the Hawaiians to try again in the 1950s.

The early lantana project did, however, inspire much more effective weed control measures in Australia. By the 1930s, the South American moth *Cactoblastis cactorum* was well on the way to clearing Australian rangelands of prickly pear cactus (genus *Opuntia*). This became the textbook case for weed control, akin to cottony-cushion scale for insect control. Australian officials next imported insects to fight another range weed, St. John's-wort (*Hypericum perforatum*), a European plant toxic to domestic animals.[17]

The weeds that troubled the Australians also infested cattle and sheep grazing areas in North America. St. John's-wort contaminated rangeland from British Columbia to northern California, where it was known as Klamath weed or goatweed. Prickly pear plagued ranches throughout the southwestern United States. Harry Smith followed the Australian program with interest, but livestock owners did not want to wait for results from down under. Chemicals were too expensive and otherwise impractical for weed control over such wide areas. The concern was not that chemicals could be hazardous, only that they did not work. In 1931, an organization of wool producers urged the university to introduce the insects being tried in Australia against St. John's-wort. Smith asked USDA officials if they could approve such a project.

The deliberate introduction of plant-feeding insects struck a sensitive nerve in the USDA. This was the antithesis of plant quarantine, which aimed to keep new phytophagous species from reaching American shores and becoming pests. Fearing that the insects would attack crop plants, chiefs of the Bureau of Entomology and the Bureau of Plant Quarantine rejected Smith's proposal. Had they trusted such a project to anyone, it would have been Smith, whose authority to import entomophagous insects was unique among all state experiment station scientists.[18]

The issue arose again in the late 1930s after a cattle rancher bought Santa Cruz Island, the largest of the Channel Islands off the coast near Santa Barbara. Joined at times by a handful of ranchers in other states, he spent two decades lobbying the USDA to introduce *Cactoblastis* to the United States or at least let Smith try it against prickly pear on the island. Although *Cactoblastis* controlled North American cacti in Australia, the moth came from South America and did not occur in the north. Smith had no desire to risk *Cactoblastis,* given the commercial value of ornamental cactus and, in particular, the occasional use of prickly pear as cattle fodder during times of drought. He did not oppose the introduction on environmentalist grounds. Aesthetic or ecological damage to cactus in the wild was not an issue.

Smith allowed that the USDA could approve *Cactoblastis* on the basis of greatest overall good, though it might force growers of ornamental cactus to control the moth with insecticides. The USDA still refused, but permitted Smith to introduce native North American cactus feeders to Santa Cruz Island, beginning in 1940. The agreement strictly excluded *Cactoblastis,* for fear that it could fly the twenty miles to the mainland. A Mexican mealybug, established in the 1950s without its own natural enemies, greatly reduced the cactus problem on the island.[19]

Weed control became a permanent part of Smith's program when he finally got his chance at the Klamath weed during World War II. The old USDA bureau chiefs were gone. Smith's friend and protégé Curtis P. Clausen now headed federal work in biological control. Clausen was willing to consider a weed project, provided that safety was assured. The introduction of an insect that became a crop pest would be a defeat for the agricultural ideal Smith and Clausen shared, and a public relations disaster for biological control.

The Australian introductions looked promising, and had produced no ill effects. Prior to release, each insect species had to pass a series of tests to show that, even when starving, it would not attack desirable plants. Smith and Clausen arranged a cooperative project in 1944. One man from the university and one from the USDA, working out of the Albany lab, would introduce insects against *H. perforatum* after starvation tests with some plant groups that the Australians had not checked.[20]

Smith made another rare exception to his usual policy of hiring people trained in California. Since biological control received so little attention in other states, and only California offered specific training in the field, Smith trusted only his own people to make the method work. In this case, however, he wanted someone with a background in plant ecology who also respected biological control. Carl B. Huffaker was uniquely qualified, as Steinhaus was for insect pathology. Both had studied at Ohio State under entomologist Alvah Peterson. Peterson made sure his students knew the value of natural enemies, although he never had the resources to develop a practical program of biological control in Ohio. If Smith must hire from outside, a Peterson student was the best he could get. Still he took Huffaker with trepidation.[21]

Smith also wanted the USDA collaborator to be someone he trusted. Recent budget cuts had severely depleted Clausen's Division of Foreign Parasite Introduction. One of the victims was James K. Holloway, whom Smith knew and found acceptable. Clausen rehired Holloway and stationed him at Albany to work on Klamath weed.[22]

War prevented direct importations from Europe, where the weed had originated. Smith arranged for his Australian friends to send their insects via U.S. Army aircraft. With the help of county agricultural commissioners in northern California, Holloway handled screening and release, while Huffaker followed the results in the field. The leaf beetles *Chrysolina hyperici* and *Chrysolina quadrigemina* spread slowly, but their effect was all that Smith and his staff could want. Over the next several years the beetles reduced the toxic weed to a rarity, and restored the rangeland to productivity.[23]

Huffaker espoused a modified version of Frederic E. Clements's climax theory of plant ecology. According to Clements, climate determined the development and eventual steady state of a plant community. He considered the community an organism in itself. In

1939, Clements and insect ecologist Victor E. Shelford wrote a book that combined their respective views of plants and animals. The authors cited the rangelands of northern California as an example in which manmade disturbance, in this case overgrazing by domestic animals, had disrupted the climax. Clements and Shelford did not mention the Klamath weed, but did discuss the proper constituents of a stable community in the region. Huffaker may have taken his cue from range management specialist Arthur W. Sampson, who had applied Clements's succession theory to grazing problems. As a member of the Berkeley forestry department since 1922, Sampson was the university's primary student of Klamath weed before the biological control project began.[24]

Those trying to control the weed with insects argued that other factors besides human activity could alter the outcome of ecological succession in a given climatic zone. Frank Wilson of Australia, who conducted the search for European insects before the war, complained that plant ecologists had ignored the role insects played in determining the composition of plant communities. While he accepted competition among plant species as a primary mechanism, Wilson held that even minor damage by insects could make a particular plant a poorer competitor, and thus keep its numbers lower than if the insects had never arrived. Control of *H. perforatum* depended on giving other, more desirable plants a competitive advantage over the weed.[25]

Huffaker concluded that the new beetles had done just that in California. Other plants were replacing Klamath weed on the range. Whereas Clements would have considered a single, undisturbed, climate-determined community best, Huffaker maintained the agricultural viewpoint of his colleagues in biological control. Like many ecologists, he discarded the superorganism concept while retaining some ideas rooted in Clements's work.[26] The "best" plant mix, according to Huffaker, was the one that served livestock interests. Rather than restore perennial grasses that had prevailed before overgrazing opened the way for weeds, ranchers wanted to encourage annuals that provided better feed. Like Smith, Huffaker intended that biological control should serve California agriculture. His purpose was not to restore a pristine, natural ideal. The *Chrysolina* beetles fit into a program of range management with carefully controlled grazing to preserve the most economically desirable, though somewhat artificial, community. In any case, the strict climax view failed to account for the arrival, by chance or design, of the weed or the insects. Each remained a permanent part of the community.

To those who still doubted that insects did much to determine plant numbers, Huffaker argued that the effect could be quite large and yet go unnoticed. An observer who did not know the history of the Klamath weed might easily conclude that the beetles had very little to do with its distribution. The amount of damage they inflicted on the plants seemed inconsequential, but before-and-after pictures showed that the insects made all the difference.

Huffaker also worked weed control into the density-dependence theory championed by Harry Smith, although that theory had not governed the project. Smith held that insect populations were kept within relatively narrow upper and lower bounds by causes of death that took larger shares of larger populations, but relaxed to kill smaller shares of smaller populations. In principle, the concept applied to plants as well. After the fact, and after he became interested in population dynamics through his association with Smith, Huffaker cited the *Chrysolina* beetles' action as responsive to the density of their food source. Having made Klamath weed rare, the beetles themselves became scarce and caused only a small, almost unnoticeable percentage of plant deaths. Yet if the weed's density increased, the beetles would increase their share and drive numbers back down.[27]

After his initial success with Klamath weed, Huffaker took on other weed control projects and also moved into insect and mite control. As he did so, he became the principal California spokesman for density-dependence theory. Huffaker held that only density-dependent factors could regulate population size. In the long run, such factors must come into play even in systems frequently disturbed by forces independent of density.

As he surveyed the attacks, rebuttals, and rejoinders of theoretical ecologists in the 1950s, Huffaker found their battle to be mostly one of semantics and emphasis. Each camp, when pressed, accepted most of the basic principles of the other, but often gave them new

names. Critics of Smith and of A. J. Nicholson, the dogmatic
Australian proponent of density dependence, did not deny the
accomplishments of biological control. Those who emphasized natural
enemies, on the other hand, came to realize that climate not only set
conditions for biotic factors but could dominate the picture and, in
some cases, even preclude biological control.

The logic of density dependence and the conspicuous effect of
environmental disturbance each had its place in any comprehensive
theory of populations. Huffaker wrote that ecologists should stand
back from their sea of new terms and their narrow focus on this or
that insect. If the combatants would stop trying to make each other
look absurd, they would realize how much they had in common.[28]

To some extent, they did stop fighting and finally start doing
research with actual organisms to try and see what was really going
on. Huffaker and some of his colleagues took part, as the University
of California increased its emphasis on basic research even in
agricultural science programs. In the 1950s, Huffaker began a series
of laboratory studies of predator-prey interactions in mite populations.
Even opponents of the Nicholson-Smith school, of which Huffaker
counted himself a member, praised his experiments as among the best
to explore the concepts of population equilibria.[29]

Population dynamics theory still had little impact on the practice
of biological control in California. As Smith's theoretical papers
between 1929 and 1939 had done, Huffaker's attempts to synthesize
climax ecology, density dependence, and environmental disturbance
defended existing practices. A few conspicuous successes—the
venerable vedalia always among them—appeared to support the
theoretical framework. A larger number of failures did not actually
refute it. Practice remained more important than theory, though
probably not as much so for Huffaker as for Smith. Only insect
pathologist Edward Steinhaus put pure science above practice.

The permanent presence that people such as Huffaker and
Steinhaus brought to Berkeley, a major teaching campus, attracted
more students to the program after the war. Steinhaus started an insect
pathology course through the Berkeley entomology division. In 1947
Smith offered a biological control course there as well, spending a

semester in Berkeley to teach it. After that, Richard L. Doutt, who had finished his Ph.D. with Smith, joined the Berkeley-Albany biological control unit as assistant professor and taught Smith's course. As graduate students completed their training, several joined the academic staff at Riverside or at the new northern location. Some returned after brief stints in state or federal government service.

Even as Smith and his colleagues diversified their work with natural enemies, the new synthetic organic insecticides made their bid to change the entire pest control picture. Although biological control had reduced some pest problems in citrus prior to World War II, others remained intractable. Growers turned increasingly to chemicals. Oil sprays replaced the dangerous practice of cyanide fumigation. DDT gained a reputation as a miracle chemical from its effect on insect-borne diseases in the war effort.

University of California entomologists began aiming the compound at agricultural pests. DDT strikingly demonstrated the importance of biological control in citrus. Scattered test plots in the San Joaquin Valley in 1944 and 1945 indicated that DDT could control citrus thrips and citricola scale. For the 1946 season, Riverside Division of Entomology Chairman Alfred Boyce and his associates recommended DDT, expecting that growers were going to use the highly publicized material anyway.

Treatment of three-fourths of the citrus in the area caused the biggest outbreak of cottony-cushion scale since 1889. To meet the emergency and restore the vedalia population in the valley, biological control workers collected the beetles wherever they could. Growers paid a dollar apiece for vedalia beetles when private insectaries could supply them. Apparently, the devastating effect of the poison on the predator had gone unnoticed in test plots because beetles quickly moved in from untreated areas. Blanketing the valley with DDT in 1946 left few such areas.[30]

In various crops, DDT used for one insect frequently led to

outbreaks of others, often species that had never been pests before. Usually this could be traced to the death of natural enemies, although the chemical appeared to benefit some phytophagous species, particularly mites, in other ways. Although it remained the symbol of a new age of pesticides, DDT was an unqualified failure in citrus except as a reminder that most pests were made, not born.

Chemical entomologists often responded to such a failure with another chemical. Boyce touted a new miracle insecticide, parathion. This organophosphorus compound could kill cottony-cushion scale, among other pests, seemingly making any loss of vedalia beetles moot. Boyce believed that the time of prescription pest control was at hand. With such new tools, anything was possible.

Parathion, however, had its own problems. Like DDT, it selected for resistance in the pest populations. Unlike DDT, parathion killed people. The university temporarily halted parathion tests after a technician died spraying plots near Riverside in 1949. Ironically, the plots were to provide fruit for experiments on the toxicity of parathion to warm-blooded animals. Boyce was confident that the technician's own carelessness had caused the accident. To prevent the entire experiment from being wasted, Boyce sprayed the plots himself a few days later. Parathion killed a Florida pest control worker the same year.[31]

In Harry Smith's division, the fear was not that the new chemicals endangered human life, but that they caused as many insect problems as they solved—and did so by disrupting biological control. Paul DeBach turned DDT into a tool for demonstrating the value of natural enemies of citrus pests. It had always been difficult to evaluate natural enemies. They could fly from test plots to control plots and spoil the experimental results. Since DDT killed many predators and parasites more readily than it killed pests, DeBach could use it to exclude the natural enemies. Treated plots became the controls in what he called the "insecticidal check method." Lower pest populations in untreated plots indicated the effect of natural enemies.[32]

The war's end unleashed widespread civilian use of DDT, but also allowed the resumption of foreign exploration for biological control. Harold Compere, by then an orange grower and pesticide user, was eager to go back to Africa, where he had made two expeditions in the 1930s. Compere sailed in 1947 to search for anything that might help the citrus industry. He spent much of his time on that never-ending problem, California red scale (*Aonidiella aurantii*). Some entomologists in South Africa claimed that cessation of insecticide treatments allowed parasites to control red scale. When Compere visited the orchards, he found the pest to be as prevalent as in California. Scale-ridden fruit might satisfy local markets in South Africa, but Compere knew from his own experience that only cosmetically perfect fruit could turn a profit in California. He now doubted that red scale would ever be controlled biologically.[33]

This was the worst trip of Compere's life. He suffered food poisoning, asthma, two hernia operations, and a debilitating rash. His wife, Joan, broke her leg badly in a fall. In addition, he objected vehemently to a side trip that Smith wanted him to take. When Compere returned to California after more than a year in Africa, his career as a foreign explorer was over. Except for a search around the United States for enemies of a grape pest in 1951, he worked on parasite systematics and morphology, and tended his orchard.[34]

The red scale project continued, despite explorer Compere's pessimism. Paul DeBach turned his full attention to red scale in 1948. Several groves had remained clean of red scale for years without insecticides. He concluded that the parasite *Aphytis chrysomphali*, under certain conditions and in the absence of chemical treatments for other pests, could control California red scale on orange trees. *A. chrysomphali* had been in California for at least half a century, but previous observers had considered the species ineffective. DeBach sought out other untreated groves and tried to persuade cooperating growers to withhold chemicals at least from small sections, so that he could investigate the effects of the parasites. He hoped to find evidence showing that some orchards would fare better untreated.

The problem of insecticidal disruption of biological control had become more acute as the citrus industry moved away from cyanide fumigation and adopted other chemical treatments for scale insects. This occurred partly because of the evolution of resistance to the gas, and partly because the old procedure was considered dangerous.

Fumigation required the raising of gastight tents over the trees, usually one or a few rows at a time. The gas dissipated quickly, allowing natural enemies to move back in after the treatment. With newer materials, larger areas were treated at one time, making it more difficult for natural enemies to recover.

County agricultural commissioners provided essential assistance to DeBach's work. Harry Smith maintained a good working relationship with these government officials, having been in the state agricultural service himself before moving to the university. Smith continued to cooperate closely with them because, as quarantine officers, the commissioners had to approve the movement of any new beneficial insects into their counties. Commissioners directed the county insectaries that mass-produced natural enemies during the 1920s and 1930s, some later than that. DeBach needed the commissioners' help because they knew where the insect problems were in their respective counties, and where he might find untreated groves. Commissioners could force treatment of an infested grove judged to be a nuisance, so DeBach depended on them to locate sufficiently isolated orchards. One spray in five years could spoil the data. In this and other ways, the biological control team worked within the agricultural establishment and aimed to serve it.[35]

To show that natural enemies were indeed responsible for low levels of red scale, DeBach refined his insecticidal test and added a "biological check method." Ants had long been known to assist honeydew-secreting insects such as aphids, mealybugs, and soft scales by driving off their natural enemies. Chemical control of ants, a relatively nonintrusive procedure using baits, facilitated biological control of honeydew producers. The ants, however, did not distinguish between these insects' parasites and those of armored scales. Red, yellow, and purple scales, all of which infested citrus, did not secrete honeydew but did benefit from the action of ants. DeBach used ants to deter parasites and cause increases of red scale in experimental plots. Comparison of trees or branches from which parasites were excluded to those where parasites had free access armed him with evidence that growers could rely more on biological control.[36]

DeBach also spearheaded a campaign to raise the standing of

biological control in the entomological profession. Such a campaign was sorely needed. Although most applied entomologists already emphasized chemicals before the war, DDT launched enthusiasm to new heights. The president of the American Association of Economic Entomologists (AAEE) for 1946 declared that even widespread pests such as the housefly could be eradicated completely. "This is not a fantastic dream," he wrote, "but is something that is almost certain to happen."[37] Another group of enthusiasts paused only briefly to consider the possible effects on beneficial insects, then confidently added, "It may be that DDT will have to prove good enough to clean up all insects, to make the predators unnecessary." Such power was indeed a dream. Most economic entomologists were prepared to believe that it had come true.[38]

As early as 1945, a few lonely voices predicted that the new broad-spectrum insecticides might do more harm than good, by killing natural enemies. Harry Smith was not among the public naysayers, although Michelbacher of Berkeley was. Smith had already warned about the evolution of resistance, but that would only make a chemical ineffective, not necessarily damaging. Having spent most of his career playing down others' excessive claims for biological control, Smith was not going to speak before the evidence was in.[39] As chemically induced upsets multiplied, however, his protégé DeBach joined a small but growing chorus calling for ecologically responsible pest control. DeBach urged any experiment station entomologist interested in natural enemies to lobby for the creation of a biological control section within the AAEE, whose meetings had become little more than pesticide revues.

DeBach found an unlikely ally in Alfred Boyce, AAEE president for 1949. Although Boyce confidently touted insecticides, he had seen the effects of natural enemies as well. He worked alongside Smith at Riverside, where the two were good friends. When DeBach took votes for a committee to draw up a section charter, Boyce's name appeared on one ballot—in the unmistakable handwriting of Harry Smith. Curtis Clausen, DeBach, and Smith led the voting. With Boyce's backing, not only did biological control get a section approved at the 1949 meeting, but Clausen of the USDA was elected president of the association. At the next annual meeting, DeBach chaired a biological control symposium, in which citrus entomology was well represented.[40]

Smith had long wanted to harness Boyce's seemingly inexhaustible energy to biological control. Boyce came to Riverside in 1927, for what was supposed to be a one-year stay in the midst of his doctoral program at Cornell University. Instead, he finished his Ph.D. with entomology professor Henry J. Quayle at the Citrus Experiment Station, and plunged into chemical control of citrus and walnut insects. Boyce's work habits prompted remarks that everyone else at the station would have to pick up the pace so as not to appear lazy.[41]

In 1932, Smith approached him about becoming an explorer. The offer tempted Boyce, whose wanderlust had already led him through four years in the merchant marine. Deciding that he could never have both a family and an exploring career, he declined. Boyce succeeded Quayle as chairman of the Division of Entomology in 1941, and guided it through dizzying postwar expansion. As Boyce prepared to take a Fulbright fellowship in India for 1951, the biological control chief came calling again.

Smith asked Boyce to forego the fellowship and spend 1951 searching for parasites of the olive scale (*Parlatoria oleae*), an armored scale pest of numerous plants. An olive scarred by even one scale would not sell, so northern California's olive industry required very thorough control. Smith wanted collections throughout the native range of the insect, from Spain to Thailand. Compere's last trip had not gone well, and Smith had no one else prepared to make the dangerous journey through politically volatile regions. Boyce took the assignment and brought his family along. His wife, Janet Mabry Boyce, served as an unpaid assistant on the trip. She had earned a Ph.D. under Smith, but nepotism rules had prevented her from pursuing a professional career in entomology at Riverside.[42]

The intrepid couple used whatever transportation was available and cheerfully braved disease and civil strife across three continents. Compere, who insisted on taking a car to Africa, had grumbled constantly about traveling conditions and bureaucratic obstacles to crossing borders. The Boyces sent to Albany any parasitized olive scale they found, following Compere's "Bring 'em back alive" precept more faithfully than he did. The project occupied the Berkeley-Albany biological control group for the next fifteen years. Shipments from Iran produced the parasite *Aphytis paramaculicornis,* which greatly reduced olive scale populations as it spread through the orchards in California. Because of the industry's exacting requirements, this control was not enough. Later explorers continued to look for olive scale when other trips happened to bring them through its range.[43]

Boyce played no part in releasing the parasites or following up on their effects. His future lay in neither insecticide nor parasite research, but in administration. Upon his return to Riverside, Boyce was named director of the Citrus Experiment Station, a post he held to retirement in 1968. The olive scale project, the last major effort Smith launched, eventually became one of the classic success stories of biological control. Along the way, olive scale and its parasites were caught up in the same personality conflicts and scientific issues as California red scale on citrus. Distinct species that lay hidden in traditional taxonomy made themselves known through biological differences crucial to successful control. Decades after the fact, Boyce claimed that others had stolen some of the credit for olive scale control.

Both projects focused attention on the parasitic genus *Aphytis.* Just as Paul DeBach was uncovering the effect of *A. chrysomphali* on red scale in untreated groves, new parasites in the genus arrived from China. Civil war had prevented thorough exploration of that country prior to the 1950s. Explorers searched worldwide, but China was believed to be the origin of both California red scale and purple scale (*Lepidosaphes beckii*). Although purple scale was the lesser pest, no effective parasites attacked it in California. DeBach found that treatments for purple scale often obstructed biological control of red scale by killing *A. chrysomphali.*[44]

After World War II, an American entomologist living in China collected parasites for the University of California. Stanley E. Flanders, who screened foreign material at Riverside, reared several *Aphytis* stocks and designated them by letter. He did not know whether they would prove to be old forms tried before, new species, sibling species, or biological races. Experience showed that any of these categories could yield something useful. Systematists might never be able to give names to every biologically significant form, but pest control would not wait. Agriculture overrode other considerations.

The observation that the parasites attacked only armored scale insects was enough to warrant trial in the orchards. What Flanders called *Aphytis* "A" from Chinese shipments on red scale, and *Aphytis* "X" on purple scale, were already in the field by the time Harold Compere described them as *A. lingnanensis* and *A. lepidosaphes,* respectively.[45] *Aphytis* "D," an olive scale parasite that a correspondent sent from Egypt in 1949, was established in northern California before Richard Doutt identified it as *A. maculicornis.*

At this time, the phrase "sibling species" entered the California literature. Although Flanders and others had recognized the existence of morphologically indistinguishable species since the 1930s, the importance of the phenomenon became widely accepted only after the culmination of the evolutionary synthesis in the 1940s. In 1953 the ornithologist Ernst Mayr, who had named the concept "sibling species," published a manual for systematics with two Berkeley entomologists, E. Gorton Linsley and Robert L. Usinger.

The Albany biological control staff, obviously acquainted with the latter two authors, adopted the principles espoused in the book and immediately applied them to olive scale parasites. Parasites that Boyce sent from Iran, India, and Spain, all originally identified as *A. maculicornis,* failed to interbreed with each other or with the form already in the orchards. Doutt and a student called these sibling species. Only the one from Iran reduced olive scale populations in California. This was later distinguished with a new species name, *A. paramaculicornis.*[46]

Meanwhile, DeBach found that the newly introduced *A. lingnanensis* replaced *A. chrysomphali* throughout most of the southern California citrus belt and improved the control of red scale. The purple scale parasite *A. lepidosaphes* reduced the need for insecticides in some places, and thus also aided red scale control. Control remained inadequate in many areas, however, especially those with harsh summers. By the mid-1950s, project leaders were doing their own exploration. DeBach went to hot, dry areas of Asia and found *Aphytis melinus,* a parasite very similar to the others in its attack on red scale. *A. melinus* displaced *lingnanensis* in inland areas of southern California. Once again, control improved. *A. lingnanensis* continued to hold sway on the coast.[47]

In work incidental to red scale exploration in 1957, DeBach collected parasitized olive scale in Pakistan and sent it to Albany. He described the finding as serendipitous. Having reached an altitude too high for citrus, he noticed some fruit trees infested with olive scale. He knew that the pest was not yet completely controlled in California. From DeBach's shipment, Doutt reared a parasite that he named *Coccophagoides utilis*. It had appeared in Boyce's shipments from the same part of the world in 1951, but Doutt had been unable to keep a stock going in the lab. Then unnamed, the species was one Doutt hoped might turn up again someday. He suspected that the males were obligate hyperparasites, which later proved to be the case. This complication had prevented the establishment of other species before Flanders described the phenomenon in the 1930s.

Taking steps to assure propagation of both sexes after *C. utilis* arrived the second time, Doutt kept it alive. Such improvements in laboratory culture techniques bore fruit in the application of biological control. When released in the olive orchards, *C. utilis* thrived in the summers, the low season for *A. paramaculicornis*. The two parasites complemented each other so well that complete commercial control resulted.

Harry Smith died before *Aphytis melinus* and *Coccophagoides utilis* showed their value in California. His intellectual descendants used both cases to defend his old argument that introduction of more than one natural enemy often improved control and never impaired it. Each additional parasite either reduced the pest population or had no effect. Interspecific competition served only to eliminate less valuable parasitic species. Which parasite was "best" varied by season or region.[48]

The red scale and olive scale results rekindled old pleas for systematics to support biological control. The Californians tried the species later named *Aphytis paramaculicornis* only because they already suspected that forms taxonomists had not separated could behave differently in the field. *A. lingnanensis* and *A. lepidosaphes* matched specimens sent to California long before. No one tried to establish them in George Compere's day because they were mistaken for species already present—despite the old explorer's insistence that they were new. Yet another new parasite was confused with *A. lingnanensis* when both were first introduced in 1948. What was thought to be one species attacking both California red scale and Florida red scale (*Chrysomphalus aonidum*) turned out to be two parasites. DeBach named the second one *Aphytis holoxanthus* after the mistake was discovered. When *A. holoxanthus* was finally sent to Florida in 1960, it completely controlled one of the worst pests of citrus in that state. Proper institutional support for systematics, DeBach argued, could have prevented the twelve-year delay.[49]

DeBach's role in the olive scale project sparked a complaint that reverted to the unseemly roots of biological control. Federal and state authorities had fought bitterly over credit for importing the vedalia beetle and controlling cottony-cushion scale in the 1880s. Alfred Boyce wrote in 1987 that DeBach had taken full credit for *Coccophagoides utilis,* the second olive scale parasite, though Boyce had sent the same parasite from the same place six years earlier.

DeBach's writings on the subject were short sections of more general discussions on armored scale control and hardly had room to name every contributor, but they did imply that his own finding was new. Neither Boyce nor DeBach observed the parasite directly. It happened to emerge from material that each sent from Pakistan to California. While DeBach maintained that credit belonged to the explorer whose shipment resulted in establishment of the parasite, he finally acknowledged that Boyce sent *Coccophagoides* first.[50]

Boyce obviously took great pride in his contribution to biological control. The olive scale expedition, the only natural enemy work of his career, filled the longest chapter of his autobiography. Smith knew what he was doing when he tried to recruit Boyce. Boyce's ability to secure funding for the Division of Entomology was legendary among biological control workers who had to make do with less. As Boyce himself recalled, he never lacked for research money. The legislature and the university supported insecticide research liberally. Chemical companies happily added grants and supplies so that the experiment station could test new compounds. Officially, no strings were attached to such grants, but they ensured that the materials would indeed be tested.[51]

The type and amount of research done reflected available funding. Chemical companies would not have continued to support research if it were not likely to further the commercial development of their products. Biological control lacked the commercial potential of chemicals, because in the ideal case there would be no product to sell. Once introduced, the most valuable natural enemies worked perpetually, on their own. Even if they could be sold, they could not be patented.

Smith knew that appropriations were skewed toward chemicals. His program might have fared better under a more dynamic leader like Boyce. Boyce would have campaigned aggressively for increases—as Steinhaus did in insect pathology. Instead, because job opportunities were so few and financial aid was virtually nil, Smith discouraged students from majoring in biological control. When he did call for more attention to biological control, he argued that it would be cheaper than pesticides and would therefore serve the farmer's interests better.[52]

Although he hesitated to join the scramble for postwar research dollars, Smith did add new positions at Riverside and Berkeley in the last years before his retirement. Except for Steinhaus and Huffaker, Smith filled all of the jobs with men trained at the University of California, mostly his own former students. Usually they had worked as laboratory technicians. For their own projects, they often took over portions of their former bosses' work. Smith was loyal to his own people, and they returned "respect and admiration almost amounting to idolatry" for the man nearly all called "Prof Harry."[53]

Perhaps Smith's gentle nature blinded him to the danger of keeping his family of scientists closed. However, rigorous training in biological control simply was not offered in any other university. Steinhaus faced the same problem in insect pathology. He avowedly opposed institutional inbreeding, but no one else taught his field at the time. University policy later came to favor hiring from outside, but in Smith's day a chairman had a relatively free hand. The Berkeley entomology division was rather provincial as well, but at least it had a broader base of professors training the next generation.

Smith encouraged a friendly atmosphere, particularly in statewide division meetings held at Berkeley on big football weekends or at his desert cabin near Palm Springs. He would rather talk about college football than about insects, anyway. His light and friendly touch at the helm did not prevent tension from building among his protégés.[54] They began to quarrel over scientific and practical matters. The teamwork that had characterized biological control in California gradually broke down. After the beloved chairman retired, the Berkeley-Albany group yearned for independence from Riverside. The stress grew until the Department of Biological Control self-destructed in the 1960s.

No one in Smith's division was ready to succeed him in 1951, when he reached the university's mandatory retirement age of 67. Flanders was too absorbed in his own studies to take over guidance of the whole division. Compere, a high school dropout with no degree or academic rank, could not be considered. He surely would not have wanted the job anyway. Everyone else was too new.

The chairman of entomology at Berkeley, Edward O. Essig, hoped to split biological control between the two campuses and take the northern unit into his own division. Smith and Essig had been associated since their early days in the State Commission of Horticulture, but by the late 1940s Essig had little use for biological control methods. He is said to have chided students for wasting their time on what he considered to be mostly propaganda. If he could annex the division, he could also gain the upper hand in his continuing fight with Steinhaus over laboratory space on the Berkeley campus.[55]

Institutional autonomy had sustained California's biological control program through depression, war, and the revolution in pesticide technology. The division's exclusive commitment to the biological method prevented that method from fading away as it largely had in other states and in the federal government. Smith opposed Essig's bid to divide and conquer. So did Claude Hutchison, dean of the College of Agriculture, but he was due to retire a year after Smith. Smith was sure that grower organizations, county commissioners, and the state department of agriculture would also object, but he feared that Essig might take over before those organizations had a chance to stop him.[56]

Hutchison and Smith found the solution to their problem in the collapse of biological control in the USDA. Curtis Clausen, head of the Division of Foreign Parasite Introduction in the Bureau of Entomology and Plant Quarantine, announced in disgust that he would resign at the end of 1950. The position of Clausen's division had always been tenuous, its jurisdiction limited to foreign exploration. Once natural enemies cleared quarantine, they passed to the various crop-based divisions, where biological control was little understood or appreciated. Foreign Parasite Introduction had not recovered from the severe curtailment of work overseas during the war.

Clausen's division breathed one last gasp in an emergency project to control the oriental fruit fly (*Dacus dorsalis*) in Hawaii. After this pest of numerous crops was discovered in the islands in 1946, Hawaii and California agencies collaborated with the USDA to reduce the threat to mainland agriculture. A dozen explorers covered twenty-one countries and sent numerous parasitic species. Three of these parasites reduced the fruit fly population, although they never controlled it completely. The Division of Foreign Parasite Introduction initially directed both exploration and the domestic work of reception, release, and evaluation.[57] Now the division might finally attain the integrated, team approach of California.

It was not to be. As soon as the foreign phase of oriental fruit fly work was over, the USDA withdrew, over Clausen's objections. From the inception of the division, he had recommended that it be given responsibility for all aspects of biological control. Losing the fruit fly project was the last straw. After Clausen left, the bureau merged parasite introduction with bee culture, a preposterous combination although both involved beneficial insects. The leaders of the new division fared no better than Clausen had. "It is apparent to us," they complained, "that the Bureau's personnel as a whole are insecticide-minded and that the insecticide industry is exerting a tremendous influence on the Bureau even to the extent that it dictates the character of much of its research."[58]

In California, an autonomous unit pursued biological control and was somewhat protected from the wave of enthusiasm for insecticides. Upon learning that Clausen was quitting the USDA, Hutchison and Smith immediately asked him to take over when Smith retired. It was the only job that Clausen would even consider. He made his acceptance contingent on the division remaining whole and separate. The very lack of such autonomy had crippled his unit in the federal service. The arrival of a prominent, California-trained entomologist with a long career in biological control, much of it as an administrator, squelched Essig's takeover bid.[59]

Colleagues hoped that retirement would finally give Smith time to write a book on biological control, including his views on population dynamics. He kept putting it off, as he had since the 1920s. The ideas were not complete yet, or he lacked time to write, or he was easily distracted—by students, new projects, the desert cabin, or football. Smith did not take well to retirement. University pensions at that time were low. He had little money left over after raising six children. He said he had no time to write one requested paper, because he needed to get his property in shape to rent, in order to pay the bills.

Smith lived in Riverside and remained in contact with his former colleagues, but they watched the man they all revered come to a sad end. He often retreated to the rugged desert country he loved. Once a teetotaler, Smith acquired a taste for wine while hunting insects in Europe in 1939, and in his last years he drank heavily. Two of his sons also became entomologists but, ironically, both went into chemical control. Sam Smith's accidental death from pesticide poisoning hit his father hard. Harry Smith died of a stroke not long afterward, on Thanksgiving Day, 1957, the day before his seventy-fourth birthday.[60]

After expanding activities statewide, Smith had talked about moving his headquarters to Berkeley, the university's main campus and the place where systemwide funding decisions were made. University administrators, however, mindful that the southern citrus industry remained the program's biggest backer, assigned Clausen to Riverside. Clausen had no experience as a teacher and, like Smith, he lacked a Ph.D. To maintain graduate advising in Riverside, Stanley Flanders was named professor of biological control in addition to his experiment station title of entomologist. Clausen also became a professor in 1955, but instruction in agricultural subjects remained very limited at Riverside during his tenure.[61]

The 1950s and early 1960s were a time of reorganization and growth in the University of California. Undergraduate instruction expanded on the northern agricultural campus at Davis, where a separate College of Agriculture opened in 1952. In that year, agricultural experiment station divisions on all campuses were upgraded to academic departments. After several years of planning, the university opened a liberal arts college adjacent to the citrus station at Riverside in 1954. By 1959, the small-college plans were scrapped. Riverside, Davis, and Santa Barbara joined Berkeley and Los Angeles as general campuses, with new campuses to be built at San Diego, Irvine, and Santa Cruz by 1965. Graduate and undergraduate curricula broadened at Riverside with the opening of a College of Agriculture in 1960, when many scientists who held only experiment station titles also became professors. Station director Boyce became dean of the college.[62]

Clausen took over a department experiencing its own growing pains. Expansion of the staff required a change in the team approach that had prevailed for most of Harry Smith's tenure. Separation of the functions of planning, exploration, and release of natural enemies became unworkable as new projects began and biological control spread to a second campus. Individuals wanted to enter new research areas without waiting for the chairman to assign them.

Huffaker, for example, yearned to move beyond weed control. He appreciated the free hand Smith gave a scientist after assigning a project, but wanted to avoid the constraint of assignments altogether.

Smith held that assignments were necessary to assure maximum benefit to agriculture. "The idea of 'academic freedom,'" he told Huffaker, "cannot, in my opinion, be rigidly adhered to in an Experiment Station."[63] Although Clausen allowed scientists to expand their work when possible, he continued to make assignments. Agricultural priorities remained intact.

The overall interdependence of the team declined. In addition to exchanging insects with entomologists in other countries, project leaders began to do their own foreign exploration. This no longer had to be coordinated for the whole department. Air transportation had already cut shipping time and increased the survival rate of insects en route. Explorers now did not need to haul cumbersome plant material around the world. By the 1950s, they could travel by air themselves and make shorter but more frequent trips than their predecessors. It became possible to ship adult parasites, rather than host insects containing parasitic larvae. This procedure made quarantine handling easier.[64]

The new chairman retained Smith's practice of light supervision. Clausen tended to require more paperwork, however, in keeping with his experience in the federal bureaucracy. He was no more aggressive in starting new projects or seeking funds, but he did obtain a share of the large legislative appropriation to deal with a new alfalfa pest that threatened both northern and southern California.[65] In the meantime, however, friction grew within the inbred department he had inherited.

Two of the old veterans played out a running feud in their respective orange orchards. Flanders acted as an idealist for biological control, while Compere took the hard-line stance of a grower who would take any measure to maximize the return on his crop. Flanders tried each year to avoid insecticides on his extensive acreage north of the Riverside campus. Compere, who regularly treated his own grove to the southwest, teased his colleague incessantly. Late in the season, proliferating red scale would make Flanders nervous, and he would finally spray. Although they disagreed, the two obviously took the interests of the citrus industry to heart, being growers themselves. The exchanges were mostly good-natured, perhaps less so after both men retired and were forced to share an office.

After years of trying, Compere's close friend DeBach finally convinced him that *Aphytis* could control red scale in some areas. Those areas did not include Compere's Arlington Heights grove, although his colleagues took some pleasure in a red scale increase that followed a spray there. Compere made good money on his fruit, but Flanders got the last laugh by selling off his property for subdivision, at great profit.[66]

Flanders and DeBach also feuded, and this was not friendly. They differed on several issues regarding red scale control (DeBach's province), but especially over the relative value of certain parasites. Flanders believed that *Casca chinensis,* an internal parasite discovered just after the turn of the century but never established in California, would solve the red scale problem. He traveled to Hong Kong in 1954 primarily to bring *Casca* back. DeBach favored the external parasites in the genus *Aphytis*. He believed *Casca* rarely if ever even attacked red scale in the field. He considered Flanders's promises to growers irresponsible. DeBach also grew tired of his colleague's propensity in later years for publishing conclusions with no supporting data. Talk among the staff held that if Flanders thought about something long enough, he came to believe he had proven it—whereupon he sent it off to a journal.[67]

While individuals quarreled in the south, rebellion brewed in the north. Clausen did not attract the affection and loyalty that had come to the exceptionally warm Harry Smith. The new chairman rarely visited Berkeley. Directed from four hundred miles away, members of the northern division felt like second-class citizens. This impression was no doubt exacerbated by the fact that, outside of biological control, north tended to look down on south. People at Berkeley always considered theirs the premier campus of the system, while Davis asserted its claim to be *the* agricultural campus. Having to report to headquarters at the comparatively tiny Riverside campus increasingly irritated the Berkeley scientists.

Clausen named Richard Doutt vice chairman to handle day-to-day operations at Berkeley-Albany. Doutt's list of grievances grew steadily. Clausen almost never came north. According to Doutt, Clausen's bureaucratic outlook led him to oppose paying students for any work that fit into their thesis research, though this was an increasingly important means of supporting graduate students. Perhaps worst of all, Clausen asked Doutt to do quarantine work on enemies of pests not established in the Bay Area. Deliberate importation of new pests, even into approved quarantine facilities, required the permission of the county agricultural commissioner, who supported Doutt's position. Doutt believed that Clausen failed to understand the importance of the long-term relationship Smith had built with the county commissioners.

In 1957 Doutt launched an attempt to break away and form a separate Berkeley department of biological control, which, he argued, already existed *de facto*. The Berkeley-Davis sphere, generally considered to cover all of the state except the handful of counties south of the Tehachapi Mountains, included nearly all major California crops and pests. Berkeley claimed to have a superior location for biological control work. The Riverside staff opposed the move. Harry Smith, two months before his death, apparently worked behind the scenes to prevent the dissolution of the unit he had built. University administrators turned Doutt down, but northern secessionist sentiment remained below the surface, ready to emerge again a few years later. Separation was probably inevitable amid the systemwide trend toward more independent campuses.[68]

Despite intercampus tensions, the entomology departments cooperated with the Department of Biological Control to fight the spotted alfalfa aphid (*Therioaphis trifolii*). A team led by Vernon M. Stern (entomology, Riverside), Ray F. Smith (entomology, Berkeley—no relation to Harry Smith), Kenneth S. Hagen (biological control, Berkeley), and Robert van den Bosch (biological control, Riverside) studied the effects of different pesticides and natural enemies on alfalfa in a variety of California environmental conditions. Natural enemies, including nonspecific ones already present and specific parasites introduced for the project, helped a great deal, but did not control the aphid completely. However, carefully timed applications of relatively selective insecticides, rather than broad-spectrum poisons such as DDT, resulted in excellent control at relatively low cost.

Stern, Smith, Hagen, and van den Bosch reported their findings in 1959. Their lead paper, entitled "The Integrated Control Concept," outlined a philosophy of pest control with a new appreciation for the value of natural enemies; a stern attack on calendar-driven, insurance treatments with dangerous and often counterproductive chemicals; and a reminder that insecticides, properly used, were still essential to modern agriculture. The authors emphasized the permanence of biological control where it worked, compared to the temporary effect of chemicals. The paper was cited frequently as the beginning of a comprehensive, ecologically aware approach that took the political catchphrase "integrated pest management."[69]

The paper had unprecedented influence. The timing was right, given the growth of pesticide use and its complications. Only four years earlier, a former director of biological control in Fiji had declared biological and chemical controls absolutely incompatible.[70] Yet the precepts of Stern et al. were nothing new. Integration of chemical and biological control methods had been a goal of practitioners of the latter for the better part of a century. The approach had matured in the University of California, but not there alone, in the 1940s. What integrated control gained in the 1950s was a name, and a broader perception of need.

"The Integrated Control Concept" followed most directly from A. E. Michelbacher's work at Berkeley since the 1930s. Alfalfa, a major feed crop in both north and south, was an especially appropriate setting for the joint project. After World War II, his former graduate student Ray Smith adopted Michelbacher's philosophy of minimal chemical disturbance. Beginning in 1949, Smith arranged for groups of alfalfa growers to hire graduate students to tell them when to spray or, more important, when not to, in order to preserve natural enemies. Michelbacher had suggested that private consultants could best assess conditions in the field. Farmers lacked the expertise and the time to follow the insects so closely, but also lacked the money to hire advisers individually.

"Supervised control," as Ray Smith called it, was an old concept. County agricultural commissioners in southern California had given such advice to citrus growers for many years. But the use of private consultants was new. It reassured growers that they were taking action, even if they were not spraying. The advisers reduced insecticide use without increasing crop damage. The effectiveness of parasites so impressed Kenneth Hagen, the first student consultant, that he switched to biological control and eventually joined the Albany faculty. Robert van den Bosch, Ray Smith's first doctoral student, became Harry Smith's last hire and initiated biological control work on field crops at Riverside. Supervised control evolved into a licensed profession known as "pest control advisers" for both field and orchard crops. The use of private consultants began to convert farmers from total dependence on chemicals, although most continued to spray early and often.[71]

At the same time that supervised control was developing in the north, Harry Smith put a newly minted Ph.D., Blair R. Bartlett, in charge of a formal project at Riverside on the effects of chemicals on biological controls. Bartlett compiled examples of pest outbreaks caused by insecticides. He urged growers to revise the timing, amount, and choice of insecticide for pests that could not be controlled any other way, in order to conserve natural enemies of other pests. Bartlett called this "integrated control," the term his colleagues finally adopted. With or without the name, this was the approach his friend DeBach already advocated for red and purple scale on citrus, and Michelbacher and Ray Smith for alfalfa insects.[72]

A group of Canadian entomologists, led by A. D. Pickett, articulated the same position independently of the Californians after watching broad-spectrum chemicals make pest problems worse rather than better in Nova Scotia apple orchards. These observations were well along before DDT even entered the picture. By the early 1950s, years before Stern et al. described "the integrated control concept," Pickett and his associates weaned growers from the most devastating fungicides and insecticides while maintaining satisfactory crops. Unlike his counterparts in California, Pickett suggested that the profit motive driving modern agriculture and, especially, its chemical component, was inappropriate. Feeding the world, he wrote in 1949, must take precedence.[73]

The main reason more entomologists turned to a less intensive chemical program in the late 1950s and early 1960s had nothing to do with ethical or environmental concerns. Advocates of integrated control simply found that chemical technology failed to do its job in agriculture. Rapid development of resistance to one chemical after another, and frequent outbreaks of secondary pests following sprays for a key pest, forced entomologists to broaden their outlook.[74] They recognized that pesticide salesmen encouraged too much chemical consumption through frequent contact and a sympathetic tone with farmers.

But in this period even the leaders of biological control accused the chemical industry only of shortsightedness, not dishonesty. Supervised control might counteract the ubiquitous presence of salesmen. Advisers with greater knowledge of ecological complexity and without a vested interest in insecticide sales could also have direct, and fairly frequent, contact with farmers.

Natural enemies worked largely unseen, whereas a host of insect corpses testified to the efficacy of (and apparent need for) a spray treatment. Integrated control faced the difficult hurdle of convincing farmers to tolerate insects in their fields. Although the Canadians urged acceptance of superficial damage to crops, the Californians did not. Instead, they worked to persuade farmers that most of the insects flying in the fields were harmless or even beneficial, and that the mere presence of a pest species did not necessarily portend economic damage.[75]

In alfalfa, grown as animal fodder, cosmetic damage was not an issue. For crops where appearance mattered, such as citrus or olives, the Department of Biological Control did not advocate reducing standards, but sought cheaper and more lasting ways to meet those standards. This often required lower insect populations than biological control achieved. The scientists took that as a challenge, not a cause for complaint. For all members of the department before the 1960s, and for most even then, retreating from the picture-perfect produce on which California had built its market was out of the question. All agreed that insecticides were necessary.

Concern for human health or for the environment did not yet intrude on the agricultural ideal. Biological control workers believed, as did their counterparts in chemical control, that unsafe pesticides had already been taken off the market and that proper use of modern compounds posed no hazard. The question had arisen enough times before the 1960s that safeguards were already in place. The entomological profession assembled plenty of evidence to defend this position by 1962, when Rachel Carson brought unprecedented national attention to the dangers of insecticides through her book *Silent Spring*.

Besides highlighting the threat chemicals posed to human health, Carson bemoaned the destruction of birds, fish, and other wildlife. Economic entomologists knew virtually nothing about this subject. Agricultural science was not concerned with plants and animals beyond the farm. Uncovering the effects of chemical inundation on wildlife fell to workers in nonagricultural agencies such as Carson's former employer, the U.S. Department of the Interior. Unable to defend themselves against evidence of environmental damage, entomologists largely ignored it and focused on direct toxicity to humans. Accumulating evidence of environmental damage—and the authority of scientists other than economic entomologists—ultimately led to the federal government ban on DDT in 1972.[76]

The few economic entomologists who helped Carson gather information, including Paul DeBach of California, faulted chemicals not for their broad environmental effects, but for their frequent failure as agricultural tools. Biological control workers agreed that industrial farming, to which they were committed, was an artificial system that would not survive a sentimental return to "nature's way." Modern agriculture was their experience, and their living. They had little use for the highly charged environmentalist movement that emerged in the 1960s with an antichemical mission—"eco-freak types," as one member of the Albany staff still called them two decades later. Harold Compere went to his grave defending the spotless orange and the chemicals he used to grow it.[77]

One California entomologist did associate openly with this movement in later years. In 1978, Robert van den Bosch published *The Pesticide Conspiracy,* a largely undocumented attack on the chemical industry, the land grant universities, and the USDA. Van den Bosch himself called the book "a barn burning polemic." But even he, colleagues recalled, privately expressed reservations about the company he was keeping, and he admitted that chemicals were sometimes necessary. He called for integrated control, not for total reliance on natural enemies. In general, biological control workers were at pains to distance themselves from anti-insecticide extremists. DeBach agreed that salesmen often scared growers into spraying unnecessarily, but he and others considered van den Bosch's angry assault on the entire industry to be counterproductive.[78]

By then a few others, including DeBach, were calling for relaxation of cosmetic standards to broaden the applicability of biological control. When DeBach made such a remark regarding red scale at a Riverside conference in 1970, an entomologist from Sunkist Growers, Inc., the dominant cooperative, objected vehemently. The position of Sunkist was that California's place in the market depended on a scale-free standard. The industry that had supported the biological control program for so many years was still not ready for a change that would allow greater reliance on natural enemies.[79]

Ironically, Florida orange growers took greater advantage of biological control than did their counterparts in the state that had sponsored the most research. Soon after the new chemicals appeared in the late 1940s, Florida entomologists realized that problems were getting worse, not better. At about the same time, the advent of frozen concentrate changed the pest control picture. Florida sold an increasing majority of its crop—more than 90 percent by 1971—as processed juice. Cosmetic standards did not matter, because consumers never saw the harmless scars on the rind of the fruit. Growers could tolerate higher pest populations than in California. Reduced pesticide use allowed natural enemies to work.[80]

Rachel Carson offered biological control as an alternative to the menace she portrayed pesticides to be. Even before she published *Silent Spring,* increasing numbers of entomologists were addressing the failure of DDT and other chemicals to provide the miraculous returns once promised. In the 1960s, after *Silent Spring,* biological control became politically correct. Instead of taking a leading role, California's Department of Biological Control nearly collapsed during the decade, and did lose its unique institutional status. Personality conflicts that had been building within the group for years finally tore it apart.

Curtis Clausen surprised the staff by stepping down from the department chair in the spring of 1959, one year short of mandatory retirement age. Whether residual tension from the Berkeley secession attempt affected his decision is unknown. The choice of a replacement was even more startling than Clausen's resignation. Administrators named Charles A. Fleschner of Riverside. An associate entomologist outranked by several members of the department, Fleschner had not campaigned for the post. Exploring in Asia at the time, he was as surprised as anyone to learn that Clausen was leaving. The northern group had expected the chair to alternate between north and south, with the next chairman at Berkeley. If the chairman was to be from Riverside, most expected DeBach to get the job.[81]

Before completing his Ph.D. in 1948, Fleschner had worked as DeBach's technician. Taking over one of DeBach's projects, mite control, Fleschner devised a new technique for evaluating the natural enemies of mites. He believed that DeBach's check methods changed the environment too much to permit the assumption that loss of natural enemies caused all pest increases. Fleschner's method was removal of predators, namely other mites, by hand—all day, seven days a week. Impractical in many systems, the method did demonstrate the importance of predators in mite control on citrus and avocados.[82]

Fleschner came to the chair full of enthusiasm for biological control. Although he did not believe that natural enemies could entirely replace insecticides, he considered work with natural enemies to be the only purpose of his department. Colleagues complained that Fleschner opposed cooperation with the entomology department in both research and teaching. Edward Steinhaus, who already wanted to break away and make invertebrate pathology a separate department emphasizing basic research, got his wish over Fleschner's objections in 1960.[83]

The new chairman sought funding increases more aggressively than his predecessors. The staff complained, however, that Fleschner also wanted direct control over their research, to the point of deciding when projects should be shut down. Coming at a time when agricultural sciences trended toward greater freedom of choice in research topics, this striking change from the style of Harry Smith and Curtis Clausen especially angered the Berkeley division.

The staff also faulted Fleschner for failing to support the preparation of a biological control book. The book, planned under Clausen and continued with DeBach as editor, dragged on through squabbles over who would be included on the various chapters. Fleschner had once headed the committee planning the book, but he backed out of the writing even before becoming department chairman. The department's master work, *Biological Control of Insect Pests and Weeds,* with sixteen authors, finally appeared in 1964.[84]

More than anything else, personality conflicts brought Fleschner down. The first chairman to rise from the ranks of his colleagues, he could not muster the respect they accorded Smith and Clausen. Fleschner's style rubbed all at Berkeley and several at Riverside the wrong way, but no one grumbled more than Robert van den Bosch. In his book, van den Bosch suggested that much of his conflict with superiors at Riverside had been due to their efforts to make him switch from field crop research to citrus.[85] The abrasive, profane, and zealous van den Bosch led a campaign of character assassination. Campus administrators tried repeatedly—and in vain—to quiet him down.

At a statewide department meeting in 1961, the majority opposed to Fleschner suddenly stood and asked him to resign. When he refused, they pressed their demands with Boyce and University Dean of Agriculture Daniel G. Aldrich. Fleschner decided to give up the fight. Boyce and Aldrich, who had hired him in the first place, rejected his resignation. They refused to let a hothead like van den Bosch tell them whom to place in charge of a department. But regardless of who was at fault, by this time the group could not function well under Fleschner. The sorry situation continued for nearly three more years before a broken Fleschner was finally allowed to step down.[86]

The trend toward campus independence within the University of California system, along with administrative desires to abolish departments lacking separate teaching programs, broke the stalemate. The Berkeley-Davis entomology department split in 1963. Berkeley formed a "superdepartment" of entomological sciences. It took in

Steinhaus's insect pathology program and the northern half of biological control as semiautonomous research divisions, and permanently legitimized the existing arrangement under which all teaching was administered through entomology. Biological control retained its division status and its separate location on the Gill Tract in Albany.

The sundering of the Department of Biological Control allowed Riverside to get rid of the troublesome van den Bosch by transferring him to Berkeley. Van den Bosch thrived there, although he remained as volatile as ever. He died of a heart attack at age fifty-six in 1978, just weeks after his controversial book appeared. In Riverside, where biological control remained a separate department, Fleschner finally left administration to revive his research at the end of 1963. Neither he nor his career ever fully recovered from his ordeal as chairman.[87]

An outsider was brought in to begin the healing process in the department. Donald A. Chant, a Canadian entomologist who shared Fleschner's interest in mites, took the chair. Chant was another surprising choice. Two years earlier, he had co-written a paper blasting biological control for its lack of theoretical guidance. Among the criticisms he and Canadian colleague Albert L. Turnbull made was that introducing more than one natural enemy could increase the pest populations. Chant arrived in Riverside enthusiastic for population modeling and basic research. He hoped that insect control would one day become a predictive science, particularly regarding the selection of natural enemies to import.[88]

Even amid the new push for basic research in the university, theory did not have much influence on biological control practice in California. Challenges to such orthodox positions as that favoring multiple introductions failed to sway the Californians. They proudly pointed to a long record in which no newly introduced natural enemy had ever made the target pest worse. There was still no way to tell in advance what an insect would do, to predict whether the new species would control the pest or even survive in a new environment.

Harry Smith's intellectual descendants considered it the height of absurdity to spend twenty or thirty years studying every detail before making an introduction, while farmers cried for help. Even then, one would not be able to predict the outcome. From a practical point of view, the time—perhaps the greater part of a scientist's career—would be wasted. In biological control in California, that practical outlook, nurtured by Smith, was what mattered.[89]

Coming from Canada, Chant represented a quite different approach. Canadian entomologists developed elaborate methods of modeling population interactions. Following the lead of W. Robin Thompson and some other British Commonwealth ecologists, they dismissed density-dependent regulation as either incorrect in nature or irrelevant to the artificial conditions of agriculture. They denounced the trial-and-error approach of the Californians and asserted that only long, painstaking, basic research could lead to improvement in pest control practice. The institutional structure of Canadian agricultural science strongly encouraged basic research. This ivory-tower approach retarded practical application in Canada.[90]

Chant did not stay in California long enough to make any great changes in department policy. Whatever his aspirations, he allowed that, for the time being, trial-and-error must continue. He supported the department's traditional activities in the practical search for and importation of natural enemies during his brief tenure.[91] After three years, Chant returned to Canada as a university administrator.

Insect pathologist Irvin M. Hall became chairman at Riverside in 1966. He had chosen to remain in the Department of Biological Control rather than become the only southern member of Steinhaus's Berkeley-based Department of Insect Pathology in 1960. Hall was among the few not involved in the attempted coup against Fleschner.[92]

Hall would be the last chairman. Against his wishes and those of everyone else in the Department of Biological Control, it was forcibly merged into entomology in 1969. Riverside Chancellor Ivan Hinderaker was integrating agriculture with the rest of the campus. He and his new dean of biological and agricultural sciences, W. Mack Dugger, extended that philosophy of unity to the department level, believing that the existence of two entomology departments stood in the way of cooperation. The rise of integrated control blurred the boundary between them. Dugger now sought to erase that boundary altogether. In addition, the systemwide move to eliminate departments with no separate teaching function was nearing completion. The entomology department already administered biological control courses. Biological control at Riverside remained a division in entomology until 1988, when even the divisional boundaries were abolished.[93]

Much of what had been unique about California's biological control program was gone. It was no longer a separate department. Other states, notably Florida and Texas, were increasing work on natural enemies. "Integrated pest management" received at least lip service all over the country. But California had paved the way, supporting biological control by natural enemies during the long period when other states did not. The biological method endured because its leaders fit it into the modern, industrial agricultural system taking shape in California.

Entomologists found a market for their services among farmers. California specialists in biological control found theirs particularly in one narrow sector, citrus. Grown only in a few states, citrus alone could not encourage the flowering of biological control in other agricultural colleges. Those working in California did little to spread their approach to other states. Lacking a nationwide network of biological control departments with which to exchange alumni, the California group remained inbred and isolated.

Having failed to spread, the market relationship between biological control and agriculture began to slip away. Urban sprawl started slowly pushing the citrus industry out of southern California after World War II. Citrus continued to thrive in the San Joaquin Valley, but without the citrus-town culture of the older areas. Harsher climatic extremes in the valley made some pests more difficult to control with natural enemies. More important, the industry held relatively less power over the university than in the earlier period. No one with the stature of old citrus kingpin Charles C. Teague could step in and prevent the demise of the Department of Biological Control in 1969. University administrators increasingly stressed basic research over direct application. Chemical companies could afford to take over some of the pesticide testing, because it led to commercial products.

Biological control generally produced no such product, but only a service less amenable to private enterprise than to government or academia. The public attention drawn to nonchemical methods in the 1960s might have provided a new market, had the scientists been ready to pursue it. They were not. Nor were they ready to expand their definition of biological control to include new methods that others were beginning to place under that banner.

The age of DDT in America may have ended with the federal ban in 1972, but biological control failed to seize the moment. The demise of a one-time agricultural wonder drug symbolized a national awareness of insecticide addiction. But DDT had arrived as only a more powerful drug for farmers already hooked on chemicals. It departed without fundamentally changing the pest control picture. Biological control remained throughout the period, albeit as an underutilized, underappreciated method. DDT neither pushed it off the stage nor dragged it back on.

CHAPTER 7

Conclusion:
Agribusiness and
Biological Control

Agribusiness.
This word has been used widely and pejoratively since the 1960s to describe farming as large-scale enterprise, vertically and/or horizontally integrated. According to this image, huge corporations rape both land and consumer, snatch economic control from individual farmers, and eventually destroy their livelihood. Science, improperly done, has contributed mightily to this sad state of affairs. Insecticides rank high on lists of villains.

As a less intrusive, more "natural" approach to insect pest problems, biological control received favorable publicity in the protest environment of the 1960s. Supposedly discarded in the mad rush to industrialize agriculture after the arrival of DDT, biological control would surely restore what had been lost. Science, properly done, would make it so.

Rachel Carson's *Silent Spring* (1962) and Jim Hightower's *Hard Tomatoes, Hard Times* (1973) symbolized what was wrong with American farming. A similar point of view guided a historical analysis by John H. Perkins, a participant in pest management debates in the 1970s. Carson warned of the ecological and health dangers of pesticides. Critics such as Hightower and Perkins bemoaned the replacement of small family farms by large-scale, corporate enterprises, and blamed the state and federal agricultural science establishment for research that put small farmers at a disadvantage.

Chemical insecticides have been considered a major factor in this change. They helped to establish unnecessary cosmetic standards, and to cripple farmers whose methods did not achieve those standards. Efficient use of chemicals required expensive equipment and methods of application that small farmers often could not afford. Those retaining more environmentally sound—but labor-intensive—methods found it difficult to make a living.

All of this occurred, said the critics, because agricultural science failed to serve the general public. Experiment stations and the U.S. Department of Agriculture worked only for those farmers most willing and able to adopt the new methods for all-out production, even though all-out production was not needed. Agricultural science ignored ecological principles. It promoted methods that destroyed wildlife, threw small farmers off the land, and even endangered the consumers of farm products. Thus scientists directly aided large corporations that took over farm operation or sold farm machines and chemicals. Small and large farms could benefit equally from safe practices such as biological control, but these were tossed aside because they did not fit into corporate agriculture. California entomologist Robert van den Bosch took this argument to the extreme. He accused the chemical industry of subverting the agricultural research establishment to maximize insecticide sales regardless of need.[1]

As I have tried to show, such views exaggerate both the prior importance of biological control and its twentieth-century disappearance. Even these writers, however, neither reject pesticides utterly nor depict biological control as the antithesis of modern agriculture. To do so would be to ignore the context in which the method developed most fully. It was the defenders of the chemical status quo who portrayed its critics as antimodern, linking them to all manner of unorthodox views.[2] In truth biological control survived, in the state most dominated by corporate farming. This happened neither because a heroic band of scientists held out against the evils of agricultural capitalism, nor because one unusual crop did. The citrus industry, which sustained biological control for so long, was not the exception it might seem to be. And, as Perkins pointed out, "integrated pest management," with its renewed recognition of the value of natural enemies, arose in the context of agribusiness rather than as an alternative to it.[3]

Although the typical citrus holding was very small, the industry organized itself into large-scale business. A few big operations dominated the Sunkist cooperative. Numerous small growers went along because they could share in increased efficiency. Growers marketed oranges as a corporation and controlled pests the same way. Associations collectively employed their own pest control operators or hired independent ones to spray or gas large acreage.

Growers united to support county and private insectaries that aided in biological control. The corporate voice of the industry, not the pleading of a few small growers leery of insecticides, kept natural enemy research going at the University of California. Large, modern enterprises provided test plots for biological control, just as they did for chemical control. Pesticide treatments in some orchards could disrupt biological control in others. Continued success of natural enemies required cooperation. In an outstanding example, growers in Ventura County organized a "citrus pest control district." Originally intended to eradicate red scale by cyanide fumigation, the district settled into a supervised program of greater reliance on biological control than anywhere else in the industry, deeply cutting pest control costs.[4] The network of county agricultural commissioners, so important to the development of biological control, was at the forefront of organized, modern agriculture, including its chemical component.

Not surprisingly, agricultural scientists and administrators resented being characterized as corporate minions. They believed that, by serving the farmers, they served the public. The public gained a cheap, plentiful, and—the scientists insisted—safe food supply. Of course agricultural science worked with the farmers most willing and able to change. The job of an entomologist or plant breeder was to find a cheaper way to grow more food and fiber, not to anticipate and forestall the social costs of agricultural change. Biological control scientists in California shared this attitude. If the farmer with large acreage could most easily afford to leave a portion untreated for study, then the scientist would take advantage of it. Scientists could not distinguish between helping small farming and helping large farming. Yes, farmers were the clients of agricultural science. In the scientists' eyes, it could be no other way.[5]

Even commentators as divergent as the environmentalist writer Frank Graham, Jr., and Mississippi congressman Jamie L. Whitten, author of a lengthy rebuttal to *Silent Spring,* agreed on this point: Biological methods would be the ideal if, and only if, they helped farmers to make a profit.[6]

It may be unreasonable to blame scientists for pursuing what seemed a sound ideal at the time. Their critics stand on firmer ground, however, in treating entomologists' response after the social and environmental costs of modern farming emerged. After small farmers went out of business, after birds and fish died, after pesticides polluted drinking water, most agricultural scientists—including those in biological control—still claimed that it was not their problem. Although some began to admit that a broader environmental perspective was necessary, and even that the cosmetic standards they had defended did not serve the public, biological control workers remained reluctant to join with environmentalists. To adopt a new clientele, such as the organic gardening movement, seemed to risk alienating the agricultural establishment for which the scientists had always labored.

To an agricultural scientist, relaxing cosmetic standards would have been tantamount to admitting that pest control could not do what had been asked of it. Economic entomologists considered themselves scientists, not engineers, but they carried out their work with an engineering ideal. They believed that they could make the world better by engineering solutions that overcame the forces of nature. To replace modern monoculture with complex systems more like natural ecosystems, or to replace insecticides with the parasites and predators that acted in natural ecosystems, would be to concede defeat. Never mind that abandoning monoculture and strictly cosmetic standards might increase the applicability of biological control.

A new generation of scientists, reared on Rachel Carson and the social concerns of the 1960s, could challenge the basic assumptions entomologists had made about their work. At least some new workers would enter the field with a different agenda, one that valued sparkling streams above sparkling oranges. However, no one can predict the future political clout of the environmental movement, or the extent to which it will emphasize agricultural issues. Even classical biological control can affect the environment adversely if an introduced agent is not absolutely specific in its attack on the target pest. One Hawaiian entomologist blames introduced natural enemies for the extinction of native species, including both nontarget prey and competing predators.[7]

Other trends in science cast doubt on the future of biological control. Institutions such as the University of California are pushing basic research and withdrawing from the most directly applied work that characterized the agricultural experiment station system. This policy marks a retreat from contact with farmers, leaving that to private interests possessing the resources to influence farm practices. Chemical corporations have acquired nearly every seed company in America. They now have the power to direct research and marketing toward increased use of fertilizers and pesticides. Because the products of genetic engineering can be patented, universities and their scientists are making lucrative deals that give corporations control over scientific information and research priorities, or are restricting the flow of information themselves. Universities may have merely replaced the ideal of farm profits with one of university profits.[8]

The emphasis on genetic engineering, however it is funded, may threaten biological control in another way. As the "hot" area in the life sciences, molecular biology commands funding and scientific prestige. Harry S. Smith felt the appeal of population dynamics, genetics, and the evolutionary synthesis in his time, but remained committed to agricultural service. As the power of the agricultural service ideal diminishes—and universities virtually disavow it—the lure of "pure" science may be irresistible.

Potential profits from patentable advances in molecular biology will sustain interest in application, to be sure. The lack of saleable products has doubtlessly handicapped classical biological control in a profit-driven economy. No one can secure a patent on a truly natural enemy. The product, if ideally successful, reproduces itself in the field. Research, development, and application of biological control must be supported by the farmers themselves, by taxpayers, or by nonprofit organizations. These groups may never match the marketing budgets of chemical corporations.

Nonspecialist natural enemies such as *Trichogramma,* common ladybeetles, and even praying mantises are sold to farmers and gardeners. Purchasing these insects, however, has never been proven to work in outdoor fields. Genetic engineering methods touted today as a form of "biological control" might conceivably produce a patentable parasite or predator, but making one useful in pest control is a longshot. Profit depends on continued sales. This is no doubt why the most intense recent interest in biological controls centers on microorganisms that are applied in the manner of pesticides.[9] Publicity for the old war horse *Bacillus thuringiensis* recalls the attention paid to DDT in the 1940s. Hopes for a panacea can only lead to disappointment, as has occurred with both chemical and biological methods in the past.

Definitions of biological control have expanded in part because of its political correctness. Male sterilization, genetic engineering, and other highly artificial techniques now labeled "biological control" may have very limited practical value or fail to address the environmental and sociological concerns that have made the term so popular. Diversion of funds and public attention to biotechnology weakens classical biological control, the identification and importation of specific natural enemies to subdue specific pests. Genetic engineering of plants may serve only to decrease genetic diversity in the field, aggravating the detrimental effects of monoculture.

Veterans in California reacted angrily to a new definition proposed by a National Academy of Sciences committee in 1987. The definition emphasized biotechnological techniques far removed from the ecological roots of biological control. The trendiest topic in so-called biological control, inserting the *Bacillus thuringiensis* toxin gene into crop plants, especially alarmed the Californians. They argued that making the toxin a permanent part of the plant, unresponsive to pest density, would cause the rapid evolution of resistance to what has been a useful, environmentally safe insecticide.[10]

The watering down of their specialty has dismayed the old hands, many of whom were slow to grasp the opportunity presented by *Silent Spring*. Sixteen California biological control specialists, including Carl Huffaker, Kenneth Hagen, and Paul DeBach, expressed their frustration in a foreword to the 1989 reissue of Robert van den Bosch's *The Pesticide Conspiracy*:

> The institutionalization of biological control is such that the term has been cheapened—it has lost its meaning. Everyone claims to be doing it, but the discipline is weaker now than it has ever been.[11]

Similar concerns emerge from historian Paolo Palladino's study of political and institutional wrangling over integrated pest management since the 1950s. Citing the example of a Canadian science structure that put basic research ahead of application, he argues that "pure" ecologists often deny the right of practically oriented scientists such as the Californians even to have opinions about theoretical matters. Meanwhile, high-minded theoreticians have failed to deliver improvements in pest control.[12]

Palladino suggests that entomologists "alternately donned two hats," that of the ecologist and that of the economic entomologist.[13] For the most part, he describes some individuals who filled one role, and some who filled the other. It is true that the career of many an entomologist lay mainly on one side of this divide. Some, however, really did both. Harry Smith is a clear example. Carl Huffaker, whom Palladino does discuss at some length, is another.

I have highlighted several instances in which the agricultural mission of California scientists led directly to theoretical and fundamental scientific advancements. A technological emphasis need not stifle basic science. Even more important to the present discussion, however, is the limited extent to which advances in applied biological control have depended on new pure science. The general idea of using natural enemies to control pests did not change when evolutionary theory replaced the Linnaean economy of nature. Neither did density-dependence theory and the concept of competitive exclusion alter practice, although they changed the scientific discussion pertaining to biological control. It is as one historian has described the relationship between science and clinical medicine in the 1860s and 1870s: "Theory confirmed practice; practice did not derive from theory."[14]

As historians of technology have been writing since the 1960s, practice derives mostly from practice. It is just as likely to inform theory as to be informed by theory. Even when a scientific discovery directly aided biological control, often as not the mission-oriented scientists made the discovery themselves, as in the case of obligate male hyperparasitism. For all their pleading that greater study of systematics would stimulate practical application, biological control workers have probably contributed more to systematics than systematics has to biological control. Research dollars spent on ecology and systematics will not necessarily improve practice, unless that research directly addresses agricultural systems and their environmental ramifications.

Californians pursued the practice of biological control because government and the citrus industry supported it, tangibly and intangibly. That support derived from one romantic, highly conspicuous—and quite lucky—experience. The story of the vedalia beetle and cottony-cushion scale has worn well. It remains a valuable symbol even after further accomplishments and the passing of a century.

This study began with the question of what made California different from states that did not support biological control. Exceptional success rates were not responsible; what Californians called high, the Canadians called distressingly low.[15] Above all, the vedalia episode itself set California apart. In the wake of this event, biological control acquired an institutional independence critical to its future. First in state government, and later in the university, a distinct agency devoted its full attention to the study and use of natural enemies. As any critic of government bureaucracy knows, agencies are difficult to eliminate. Thus biological control maintained its own identity amid the chemical onslaught, until 1969. Every detail of the cottony-cushion scale story contributed. Indeed, had control been as complete as that of citrophilus mealybug forty years later, cottony-cushion scale might have had less influence than it had. Flare-ups due to insecticides down through the years reminded all parties of the importance of natural enemies. Surely there are other scientific or technological events whose symbolic value alone gives them a disproportionately long historical reach.

In drawing a new picture of the relationship between science and technology, historians of technology have legitimated their own specialty, as distinct from history of science. Technology is a dynamic enterprise in itself, not merely the application of ideas developed by pure scientists.[16] Historians of agricultural science similarly find it necessary to justify their interest. As one has put it, "For historians of science, agricultural science represents a distinctly 'blue collar' phenomenon, which, like engineering, suffers from neglect partly because of its practical aspects."[17] Perhaps this is why, in a recent appraisal of the early twentieth-century dominance of the Cornell entomology department, the University of California was dismissed from comparison because of its emphasis on economic entomology.[18]

American agricultural science deserves attention not simply as a historical excuse for the training of large numbers of scientists, some of whom went on to do something "important." The institutional setting and professional identity of agricultural science have been more science than technology, yet the practical mission has a flavor of engineering. The private consultants now known as pest control advisers are a more obvious step in this direction.

Even historians interested only in pure science would do well to consider the effect that applied fields and their institutional structure have had on the development of scientific ideas. For example, professional and institutional boundaries may prove quite important in the development of systematics, if its relationship with biological control is representative. Despite paying a great deal of attention to the evolutionary synthesis, historians of biology have done surprisingly little work on twentieth-century systematics. Those who finally plow this fertile field must not neglect institutional factors.

Surely a concern for the practical—especially environmental—consequences, as opposed to purely scientific developments, underlies the recent appearance of several books on the history of pest control in America. I believe that reduced insecticide use and increasing reliance on natural enemies would be a decided improvement, even though it won't bring back the small family farm. This change would require either government regulation or economic pressure from consumers insisting that food and fiber be grown without pesticides. Relaxing cosmetic standards for produce would help, but even those scientists most dedicated to biological control have tended to avoid that issue.

Past practitioners of biological control may be heroes to the environmental movement—and villains to latter-day "squirt-gun entomologists" and their allies in industrial agriculture. Either position, however, mistakenly characterizes the likes of Harry Smith as agricultural revolutionaries. Smith and his associates participated willingly in modern, profit-driven farming, eager to make a spotless orange. A real revolutionary would say that oranges should not be spotless, as self-proclaimed radical Robert van den Bosch did in later years.[19]

Agricultural entomologists were not the only scientists newly thrust into the spotlight by the rise of environmentalism. The public has tended to equate environmentalism and ecology since the 1960s. The two fields developed along separate paths. Professional ecologists mostly studied natural systems without focusing on human disturbance. Ecologists were unprepared for the new public perception of their work. As Peter J. Bowler has recently written, countering such perceptions is part of the task of historians of these fields, who also have a changing audience due to the popularity of environmentalism.[20]

The institutional structure into which biological control was built began to change in the 1960s. The environmental movement attacked much of what had become standard in American agriculture. Public sympathy for migrant farm workers grew, and pest control became one of the issues those workers raised as they organized. "Integrated control" often gave short shrift to its biological component, but nevertheless eroded professional boundaries within economic entomology. New methods and new political situations complicated the entire pest control picture in ways beyond the scope of this study. The California biological control group, split into two parts, came only slowly to appreciate the import of these changes.

Historians have examined two alternatives to all-out chemical pest control in the 1960s and 1970s. The integrated pest management approach competed for funds and attention with the sterilization technique developed in the USDA. John Perkins has characterized the two sides as having fundamental philosophical differences over the proper relationship between man and the environment. In his view, the sterile-male technique, bent on eradication, reflected the aim of total mastery over nature, while integrated pest management was an effort to live within the constraints of nature. Paolo Palladino argues, more convincingly, that philosophical differences were exaggerated in what was really a fight for funding, driven by institutional and political factors that pitted state and federal entomologists against each other.[21] The scientists were learning to promote their services in a new way.

Biological control specialists from around the world gathered in Riverside in 1989 to mark the centennial of the control of cottony-cushion scale by its natural enemies. University of California emeritus professor Richard L. Doutt urged his assembled colleagues to take advantage of a new market. Organic farming was expanding, as was its popular appeal. City dwellers wanted to avoid insecticides on their food and would welcome a nonchemical way to control pests of their precious ornamental plants. These groups outside the old agricultural establishment, Doutt said, provided unprecedented opportunities for biological control. Doutt even called for the reestablishment of centralized authority over biological control. The university, he said, should set up a single headquarters for work throughout the state of California, preferably apart from any campus.[22]

This was the man who had led the attempt to split the Department of Biological Control in two in the 1950s. California might again become the undisputed leader if his suggestion were followed. Biological control needed the coordinated, team approach on which the old program was built.

Doutt's proposal to cultivate a new constituency drew a mixed reaction and stimulated conversation for the duration of the 1989 symposium. Many participants were still not ready to become environmentalists and give even the appearance of turning their backs on their old clients in agribusiness. The system that had become addicted to chemicals had also fostered the development of biological control. After a hundred years, practitioners of biological control still hesitated to break the bond. But some were now willing to consider it.

Notes

ABBREVIATIONS

Manuscript sources cited have been
identified by the following abbreviations:

AJN	Alexander John Nicholson Papers, Manuscript Collection 130/i, Basser Library, Australian Academy of Science, Canberra
AK	Albert Koebele Papers, Department of Entomology, California Academy of Sciences, San Francisco
BIW	Benjamin Ide Wheeler Papers, Manuscript Collection C-B 1044, Bancroft Library, University of California, Berkeley
CAF	Correspondence in possession of Charles A. Fleschner, Trinidad, California
CHR	Chester H. Rowell Papers, Manuscript Collection C-B 401, Bancroft Library, University of California, Berkeley
CSA	California State Archives, Office of the Secretary of State, Sacramento
CVR-NAL	Charles Valentine Riley Papers, National Agricultural Library, Beltsville, Maryland
CVR-SI	Charles Valentine Riley Papers, Record Unit 7076, Smithsonian Institution Archives, Washington, D.C.
EAS	Edward Arthur Steinhaus Papers, University of California, Irvine, Archives
EOE	Edward Oliver Essig Papers, Department of Entomology, California Academy of Sciences, San Francisco
EWH	Hilgard Family Papers, Manuscript Collection C-B 972, Bancroft Library, University of California, Berkeley
GCP	George C. Pardee Papers, Manuscript Collection C-B 400, Bancroft Library, University of California, Berkeley
HC	Harold Compere Papers, Department of Entomology, University of California, Riverside*
HJQ	Henry Josef Quayle Papers, Manuscript Collection 117, Special Collections, Tomàs Rivera Library, University of California, Riverside

HSH	Family papers in possession of Harriet Smith Hanson, Lafayette, California
HSS	Harry Scott Smith Papers, Department of Entomology, University of California, Riverside*
HWJ	Hiram W. Johnson Papers, Manuscript Collection C-B 581, Bancroft Library, University of California, Berkeley
PHD	Paul Hevener DeBach Papers, Department of Entomology, University of California, Riverside*
PHT	Philip Hunter Timberlake Papers, Department of Entomology, University of California, Riverside*
RG7	Records of the Bureau of Entomology and Plant Quarantine, Record Group 7, National Archives, Washington, D.C. (Entry abbreviations below)

E2	Letters Received, 1883–1908
E3	Correspondence, 1894–97
E5	Letters Sent, 1878–93
E6	Letters Sent, 1893–1908
E31	Correspondence Relating to Investigations of Gypsy and Brown-Tail Moths, 1905–8
E34	General Correspondence, 1908–24
E35	General Correspondence, 1925–34
SE8	Correspondence Relating to Bureau Programs and Plans, 1930–51
SE14	Office Files of the Chief of the Bureau of Entomology, 1880–1950
SE22	General Files on Special Investigations, 1916–50
SE62	Records and Correspondence from Division Employees on Duty in Foreign Countries, 1928–51
SE65	General File, Records of the Division of Fruit Fly Investigations, 1930–34
SE121	Records Concerning Insect Identification and Parasite Introduction, 1919–43

SEF	Stanley Ellsworth Flanders Papers, Department of Entomology, University of California, Riverside*
SGA	Field Services: Pest Control Department, Sunkist Growers, Inc., Archives, Sherman Oaks, California
UCA-CA	College of Agriculture Records, CU-20, University of California Archives, Berkeley
UCA-PF	President's File, CU-5, University of California Archives, Berkeley
UCR-BC	Biological Control Files, Department of Entomology, University of California, Riverside
UCR-DE	Personnel Files, Department of Entomology, University of California, Riverside

*Personal papers held by the Department of Entomology, University of California, Riverside, are not formally archived.

INTRODUCTION

1. Rachel Carson, *Silent Spring* (Boston: Houghton Mifflin, 1962).

2. Biological control has been defined variously as a phenomenon, a method, or an area of study. Participants such as California's Paul DeBach, a leading spokesman in the postwar period, include all three. I shall do likewise. See Paul DeBach, "The scope of biological control," in *Biological Control of Insect Pests and Weeds*, ed. Paul DeBach (New York: Reinhold, 1964), pp. 3–20.

3. Karl Escherich, *Die Angewandte Entomologie in den Vereinigten Staaten* (Berlin: Paul Parey, 1913), pp. 78, 129; anonymous review of Escherich, *Journal of Economic Entomology* 6 (1913): 430–432; Leland O. Howard, "The practical use of the insect enemies of injurious insects," *U.S. Department of Agriculture Yearbook* 1916: 273–288, on p. 287; Harry S. Smith, "Economy in insect control through the use of parasites," *California State Commission of Horticulture Bulletin* 6 (1917): 362; idem, "On some phases of insect control by the biological method," *Journal of Economic Entomology* 12 (1919): 288–292.

4. Vernon M. Stern et al., "The integrated control concept," *Hilgardia* 29 (1959): 81–101, on p. 85; Paul DeBach, *Biological Control by Natural Enemies* (London: Cambridge University Press, 1974), preface. On the use of sterile insects, see, John H. Perkins, "Edward Fred Knipling's sterile-male technique for control of the screwworm fly," *Environmental Review* 1, no. 5 (1978): 19–37.

5. Charles V. Riley, "Parasitism in insects," *Proceedings of the Entomological Society of Washington* 2 (1892): 397–431, on pp. 402–403; William Morton Wheeler, "Insect parasitism and its peculiarities," *Popular Science Monthly* 79 (1911): 431–449; idem, *The Social Insects, Their Origin and Evolution* (New York: Harcourt, Brace, 1928), p. 35; Richard L. Doutt, "The biology of parasitic Hymenoptera," *Annual Review of Entomology* 4 (1959): 161–182.

6. James C. Whorton, *Before Silent Spring: Pesticides and Public Health in Pre-DDT America* (Princeton: Princeton University Press, 1974), pp. 3–24; W. Conner Sorensen, "Brethren of the net: American entomology, 1840–1880" (Ph.D. dissertation, University of California, Davis, 1984), pp. 111–255; Margaret W. Rossiter, "The organization of the agricultural sciences," in *The Organization of Knowledge in Modern America, 1860–1920*, ed. Alexandra Oleson and John Voss (Baltimore: Johns Hopkins University Press, 1979), pp. 211–248, on pp. 220–224; Leland O. Howard, *A History of Applied Entomology (Somewhat Anecdotal)* (Washington: Smithsonian Miscellaneous Collections, 1930), pp. 10–93; Thomas R. Dunlap, *DDT: Scientists, Citizens, and Public Policy* (Princeton: Princeton University Press, 1981), pp. 18–23.

7. Whorton, *Before Silent Spring*, pp. 24–35, 69–73; Douglas Helms, "Technological methods for boll weevil control," *Agricultural History* 53 (1979): 286–299, on pp. 287–288; Ian R. Manners, "The persistent problem of the boll weevil: Pest control in principle and in practice," *Geographical Review* 69 (1979): 25–42, on pp. 31–33; Thomas R. Dunlap, "The triumph of chemical pesticides in insect control, 1890–1920," *Environmental Review* 1, no. 5 (1978): 38–47; Paolo S. A. Palladino, "Entomology and ecology: The ecology of entomology" (Ph.D. dissertation, University of Minnesota, 1989), pp. 49–63; John H. Perkins, *Insects, Experts, and the Insecticide Crisis: The Quest for New Pest Management Strategies* (New York: Plenum, 1981), pp. 11–12, 67–68. Whorton, Helms, and Manners propose the earliest dates for the chemical orientation. Dunlap stresses the 1900s and 1910s, Palladino the 1920s and 1930s, and Perkins the 1940s.

8. Sharon E. Kingsland, *Modeling Nature: Episodes in the History of Population Ecology* (Chicago: University of Chicago Press, 1985), pp. 50–53, 98–142. Kingsland set applied aspects aside to concentrate on theoretical issues in mathematical population dynamics.

9. Charles E. Rosenberg, *No Other Gods: On Science and American Social Thought* (Baltimore: Johns Hopkins University Press, 1976), chapters 8–12, especially pp. 166–168, 179–180.

Chapter 1. THE SCOURGE

1. Huang Hsing-Tsung, "Plants and insects in man's service," in *Science and Civilization in China*, vol. 6, *Biology and Biological Technololgy*, part 1, *Botany* by Joseph Needham with Lu Gwei-Djen (Cambridge: Cambridge University Press, 1986); H. C. McCook, "Ants as natural insecticides," *Proceedings of the Academy of Natural Sciences of Philadelphia* 1882: 263–271; Robert L. Usinger, "The role of Linnaeus in the advancement of entomology," *Annual Review of Entomology* 9 (1964): 1–16, on pp. 6–7; George Ordish, *The Constant Pest: A Short History of Pests and their Control* (London: Peter Davies, 1976), pp. 31–33; Vincent Köllar, *A Treatise on Insects Injurious to Gardeners, Foresters, and Farmers*, trans. J. and M. Loudon (London: W. Smith, 1840); John Curtis, *Farm Insects* (Glasgow: Blackie and Son, 1860); William Kirby and William Spence, *An Introduction to Entomology*, 4 vols. (London: Longman, 1815–1826); John Obadiah Westwood, *An Introduction to the Modern Classification of Insects, Founded on the Natural Habits and Corresponding Organization of the Different Families* (London, 1839–40).

2. W. Conner Sorensen, "Brethren of the net: American entomology, 1840–1880" (Ph.D. dissertation, University of California, Davis, 1984), pp. 170–183; idem, "The rise of government sponsored applied entomology, 1840–1870," *Agricultural History* 62, no. 2 (1988): 98–115.

3. Leland O. Howard, "Danger of importing insect pests," *Yearbook of the U.S. Department of Agriculture* 1897: 529–552, on p. 530; Asa Fitch, "Address, on our most pernicious insects," *Transactions of the New York State Agricultural Society* 20 (1860): 588–598, on pp. 596–597; idem, "Sixth report on the noxious and other insects of the state of New York," ibid., 20 (1860): 745–868, on pp. 816, 824–825 (quotation from p. 824); Benjamin D. Walsh, "Importing European parasites," *Practical Entomologist* 2 (1867): 54–55; Leland O. Howard, *A History of Applied Entomology (Somewhat Anecdotal)* (Washington: Smithsonian Miscellaneous Collections, 1930), pp. 51–52.

4. Sorensen, "Brethren of the net," pp. 277–279; [Benjamin D. Walsh], *Practical Entomologist* 2 (1866): 7–8; Charles Darwin, *On the Origin of Species by Means of Natural Selection* (London: John Murray, 1859), p. 73.

5. Benjamin D. Walsh, "Imported insects—the gooseberry sawfly," *Practical Entomologist* 1 (1866): 117–125, on pp. 118–119. On Buffon's theory, see John C. Greene, *The Death of Adam: Evolution and Its Impact on Western Thought* (Ames: Iowa State University Press, 1959), pp. 150–152.

6. Transactions of the Illinois State Horticultural Society, 1865: 65–66; ibid., 1867: 499; Charles V. Riley, "In memoriam" [of Walsh], *American Entomologist* 2 (1870): 65–68.

7. Howard, *History of Applied Entomology*, pp. 54–56, 90–92; Edwin P. Meiners, "Charles Valentine Riley," mimeographed (Columbia, Missouri: C. V. Riley Entomological Society, 1959), p. 3; Alpheus S. Packard, Jr., "Charles Valentine Riley," *Science*, n.s., 2 (1895): 745–751, on pp. 745–746; L. D. Morse, "Notice," *Annual Report of the Missouri State Board of Agriculture* 1868: iii–v.

8. Charles V. Riley, *Second Annual Report on the Noxious, Beneficial, and Other Insects of the State of Missouri* (Jefferson City: Horace Wilcox, 1870), pp. 8–13.

9. Sorensen, "Brethren of the net," pp. 284–298; Charles V. Riley, *Third Annual Report on the Noxious, Beneficial, and Other Insects of the State of Missouri* (Jefferson City: Horace Wilcox, 1871), pp. 174–175; idem, *Fifth Annual Report on the Noxious, Beneficial,*

and Other Insects of the State of Missouri (Jefferson City: Regan & Carter, 1873), pp. 174–175; idem, "Means of dealing with our insect foes," Lowell Institute Lecture, 4 January 1892, pp. 4a–5, box 2, Lectures 1868–92, CVR-NAL; idem, "Parasitic and predaceous insects in applied entomology," Insect Life 6 (1893): 130–141.

10. Sorensen, "Brethren of the net," pp. 232–255; Howard, History of Applied Entomology, pp. 78–89.

11. Howard, History of Applied Entomology, pp. 70–75; Sorensen, "Brethren of the net," pp. 148–149; Pamela M. Henson, "Evolution and taxonomy: J. H. Comstock's research school in evolutionary entomology at Cornell University, 1874–1930" (Ph.D. dissertation, University of Maryland, College Park, 1990); Riley to Eugene W. Hilgard, 23 May 1881, EWH; Comstock to Hilgard, 31 May 1881, EWH.

12. Riley to Comstock, 29 November 1878, vol. 2, E5, RG7; Riley to Albert J. Cook, 24 March 1891, vol. 57, ibid.; Riley, Third Missouri Report, p. 29; Richard L. Doutt, "The historical development of biological control," in Biological Control of Insect Pests and Weeds, ed. Paul DeBach (New York: Reinhold, 1964), pp. 21–42, on p. 31; Riley, "Parasitic and predaceous insects," pp. 131–133; idem, Fourth Annual Report on the Noxious, Beneficial, and Other Insects of the State of Missouri (Jefferson City: Regan & Edwards, 1872), pp. 38–40.

13. Charles V. Riley, Sixth Annual Report on the Noxious, Beneficial, and Other Insects of the State of Missouri (Jefferson City: Regan & Carter, 1874), pp. 7, 30–65; Leland O. Howard and William F. Fiske, "The importation into the United States of the parasites of the gipsy moth and the brown-tail moth," U.S. Bureau of Entomology Bulletin 91 (1911), p. 24; Charles V. Riley, "Report of the entomologist," Annual Report of the U.S. Commissioner of Agriculture 1883: 99–180, on pp. 108–113, 131–138; idem, "Report of the entomologist," ibid., 1884: 285–418, on p. 323; Harry C. Coppel and James W. Mertins, Biological Insect Pest Suppression (Berlin: Springer-Verlag, 1977), p. 21.

14. Charles V. Riley, "Report of the entomologist," Annual Report of the U.S. Commissioner of Agriculture 1887: 48–179, on pp. 49–50.

15. Charles V. Riley, "The scale insects of the orange in California, and particularly the Icerya or fluted scale, alias white scale, alias cottony-cushion scale, etc.," Pacific Rural Press 33 (1887): 361–364; Leland O. Howard, "The practical use of the insect enemies of injurious species," Yearbook of the U.S. Department of Agriculture 1916: 273–288, quotation from p. 274.

16. Charles V. Riley, "Report of the entomologist," Annual Report of the U.S. Commissioner of Agriculture 1878: 207–257, on pp. 208–209; idem, "Large white scale on acacias, etc.," American Entomologist 3 (1880): 20; Leland O. Howard, Fighting the Insects: The Story of an Entomologist (New York: Macmillan, 1933), p. 33; Edward O. Essig, A History of Entomology (New York: Macmillan, 1931), p. 576; Howard, History of Applied Entomology, p. 95; Harold Compere, "The red scale and its insect enemies," Hilgardia 31 (1961): 173–278, on pp. 177–178.

17. Paul DeBach, Biological Control by Natural Enemies (London: Cambridge University Press, 1974), p. 176; [Edward J. Wickson], "Parasites of scale insects," Pacific Rural Press, 27 September 1879, p. 200; Leland O. Howard, "Report on the parasites of the Coccidae in the collection of this department," Annual Report of the U.S. Commissioner of Agriculture 1880: 350–373.

18. Riley to the Secretary of State, 12 September 1885, vol. 23, E5, RG7; Riley to Daniel W. Coquillett, 7 December 1885, ibid.; Riley to Coquillett, 7 January 1890, box 27, Outgoing Correspondence and Memoranda 1886–1895, CVR-SI

19. Howard, History of Applied Entomology, p. 99, and n. 1; Riley, "Scale insects of the orange."

20. Howard to Coquillett, 4 March 1886, vol. 24, E5, RG7; Howard to Coquillett, 13 February 1886, ibid.; Howard to Koebele, 26 February 1886, ibid.; Howard to Koebele, 24 June 1886, vol. 25, ibid.; Howard to Coquillett, 6 July 1886, ibid.; Richard L. Doutt, "Vice, virtue, and the vedalia," *Bulletin of the Entomological Society of America* 4 (1958): 119-123; George Compere, "Origin of fumigation with hydrocyanic acid gas in California," *California Department of Agriculture Bulletin* 11 (1922): 438-442; Charles W. Woodworth, "Orchard fumigation," *California Agricultural Experiment Station Bulletin* 122 (1899), pp. 3-6; Riley to Koebele, 18 June 1891, AK.

21. Charles V. Riley, "Address" (to Seventh State Fruit Growers' Convention, 12 April 1887), *Biennial Report of the California State Board of Horticulture* 1885-86: 450-471; Howard to Riley, 10 October 1887, vol. 33, E5, RG7; Howard to Coquillett, 7 November 1887, ibid.; Doutt, "Historical development of biological control," p. 34; Charles V. Riley, "On the original habitat of *Icerya purchasi*," *Pacific Rural Press* 35 (1888): 425.

22. Howard to Crawford, 16 August 1887, vol. 32, E5, RG7; Riley to Crawford, 14 December 1887, vol. 34, ibid.; "Australian parasites of the fluted scale," *Pacific Rural Press* 35 (1888): 345.

23. Howard to Crawford, 16 June 1888, vol. 37, E5, RG7; Riley to Waldemar G. Klee, 13 October 1888, vol. 39, ibid.; William H. Thorpe, "The biology, post-embryonic development, and economic importance of *Cryptochaetum iceryae* (Diptera, Agromyzidae) parasitic on *Icerya purchasi* (Coccidae, Monophlebini)," *Proceedings of the Zoological Society of London* 1930: 929-971, on pp. 934-938; Crawford to Riley, 18 March 1889, box 18, E2, RG7; Howard to Koebele, 15 July 1889, vol. 44, E5, RG7; Riley to Crawford, 26 December 1890, vol. 55, ibid.

24. Harry S. Smith, "The utilization of entomophagous insects in the control of citrus pests," *Transactions of the Fourth International Congress of Entomology* 2 (1929): 191-198, on p. 192; DeBach, *Biological Control by Natural Enemies*, pp. 93-94, 99-100, 174-176.

25. Albert Koebele, "Report of a trip to Australia made under the direction of the Entomologist to investigate the natural enemies of the fluted scale," *U.S. Division of Entomology Bulletin* 21 (1890); Howard, "Practical use of insect enemies," pp. 275-276; Doutt, "Historical development of biological control," pp. 34-37; DeBach, *Biological Control by Natural Enemies*, pp. 96-97. The New Zealand source has often been neglected, probably because most accounts list only the numbers of adults in each shipment, whereas Koebele's last collection contained vastly more immature specimens. Clare F. Morales and Richard L. Hill pointed out the New Zealand connection in an unpublished presentation at the International Vedalia Symposium on Biological Control, 27-30 March 1989, Riverside, California.

26. Riley to Koebele, 5 December 1889 and 3 January 1889, AK; Riley to Coquillett, 28 December 1888, vol. 40, E5, RG7; Howard, *History of Applied Entomology*, p. 502; Compere, "Red scale and its insect enemies," p. 181.

27. Craw to Howard, 6 October 1900, box 516, SE14, RG7; Byron Martin Lelong, "Beneficial insects," *Annual Report of the California State Board of Horticulture* 1889: 260-288, on pp. 260-262; Ellwood Cooper, *Bug vs. Bug: Parasitology* (Santa Barbara, Calif., 1913), pp. 9-10; "The fluted scale," *Pacific Rural Press* 35 (1888): 110; "Damaging facts *in re* Board of Horticulture," *Rural Californian* 1895: 144-146; Riley to Coquillett, 15 July 1892, vol. 70, E5, RG7; George Compere to Cooper, 5 June 1901, HC; Riverside; G. Compere to Edward K. Carnes, 3 November 1908, HC; Craw to Howard, 28 May 1901, box 516, SE14, RG7. The structure of the agricultural bureaucracy in California is discussed in Chapter 2.

28. Doutt, "Vice, virtue, and the vedalia," pp. 122-123; Cooper, *Bug vs. Bug*, p. 6.

Chapter 2. CITRUS, GOVERNMENT, AND SCIENCE IN CALIFORNIA

1. Harry W. Lawton, "A history of citrus in southern California," in *A History of Citrus in the Riverside Area,* ed. Esther H. Klotz, Harry W. Lawton, and Joan H. Hall (Riverside, Calif.: Riverside Museum Press, 1969), pp. 6–11.

2. Tom Patterson, *A Colony for California: Riverside's First Hundred Years* (Riverside, Calif.: Press-Enterprise Co., 1971), pp. 21–39, 159–160; Kevin Starr, *Inventing the Dream: California through the Progressive Era* (New York: Oxford University Press, 1985), pp. 133–145; Harry W. Lawton and Lewis G. Weathers, "The origins of citrus research in California," in *The Citrus Industry, vol. 5: Crop Protection, Postharvest Technology, and Early History of Citrus Research in California,* ed. Walter Reuther, E. Clair Calavan, and Glenn E. Carman (Berkeley: University of California Division of Agriculture and Natural Resources, 1989), pp. 281–335, on pp. 285–289; Carey McWilliams, *Southern California Country: An Island on the Land* (New York: Duell, Sloan, and Pearce, 1946), p. 209; Oscar E. Anderson, Jr., *Refrigeration in America: A History of a New Technology and its Impact* (Princeton: Princeton University Press, 1953), pp. 155–158; Michael R. Belknap, "The era of the lemon: A history of Santa Paula, California," *California Historical Society Quarterly* 47 (1968): 113–140, on pp. 120–123; Charles C. Teague, *Fifty Years a Rancher: The Recollections of Half a Century Devoted to the Citrus and Walnut Industries of California and to Furthering the Co-operative Movement in Agriculture* (Los Angeles: Ward Ritchie Press, 1944), pp. 44–46; Lawrence J. Jelinek, *Harvest Empire: A History of California Agriculture* (San Francisco: Boyd & Fraser, 1979), p. 51.

3. Starr, *Inventing the Dream,* pp. 134–143, idem, *Americans and the California Dream, 1850–1915* (New York: Oxford University Press, 1973), pp. 201–203; Lawton and Weathers, "Origins of citrus research," pp. 285–291; Charles Howard Shinn, "Social changes in California," *Popular Science Monthly* 38 (1891): 794–803, on p. 798; J. Eliot Coit, *Citrus Fruits: An Account of the Citrus Fruit Industry with Special Reference to California Requirements and Practices and Similar Conditions* (New York: Macmillan, 1915), 10–12, 356–357; McWilliams, *Southern California Country,* pp. 211–216.

4. Roy V. Scott, *The Reluctant Farmer: The Rise of Agricultural Extension to 1914* (Urbana: University of Illinois Press, 1970); David B. Danbom, "The agricultural experiment station and professionalization: Scientists' goals for agriculture," *Agricultural History* 60, no. 2 (1986): 246–255; Coit, *Citrus Fruits,* p. 10; Lawton and Weathers, "Origins of citrus research," p. 291.

5. Pete Daniel, *Breaking the Land: The Transformation of Cotton, Tobacco, and Rice Cultures Since 1880* (Urbana: University of Illinois Press, 1985), pp. 37–61.

6. Gerald D. Nash, "The sugar beet industry and economic growth in the West," *Agricultural History* 41 (1967): 27–30; Howard Seftel, "Government regulation and the rise of the California fruit industry: The entrepreneurial attack on fruit pests, 1880–1920," *Business History Review* 59 (1985): 369–402, on pp. 375–376; Alan L. Olmstead and Paul Rhode, "An overview of California agricultural mechanization, 1870–1930," *Agricultural History* 62, no. 3 (1988): 86–112.

7. Verne A. Stadtman, *The University of California, 1868–1968* (New York: McGraw-Hill, 1970), pp. 143–147; Alan I Marcus, *Agricultural Science and the Quest for Legitimacy: Farmers, Agricultural Colleges, and Experiment Stations, 1870–1890* (Ames: Iowa State University Press, 1985); pp. 18–35; Ronald L. Nye, "Federal vs. state agricultural research policy: The case of California's Tulare experiment station, 1888–1909," *Agricultural History* 57 (1983): 436–449; Charles E. Rosenberg, *No Other Gods: On Science and American Social Thought* (Baltimore: Johns Hopkins University Press, 1976), pp. 164–165.

8. Lawton and Weathers, "Origins of citrus research," pp. 294–306; John Henry

Reed, "California Farmers' Clubs," *Pacific Rural Press* 67 (1904): 278; Starr, *Inventing the Dream*, pp. 137-140; Stadtman, *University of California*, p. 149; Herbert J. Webber, Walter Reuther, and Harry W. Lawton, "History and development of the citrus industry," in *The Citrus Industry*, vol. 1: *History, World Distribution, Botany, and Varieties*, ed. Walter Reuther, Herbert J. Webber, and Leon D. Batchelor (Berkeley: University of California Division of Agricultural Sciences, 1967), pp. 1-39, on pp. 33-34.

9. Seftel, "Government regulation," pp. 377-381; Edward O. Essig, "Official entomology in California—some comments, historical and personal," *California Department of Agriculture Bulletin* 44 (1955): 3-16; idem, *A History of Entomology* (New York: Macmillan, 1931), pp. 581-585; Gerald D. Nash, *State Government and Economic Development: A History of Administrative Policies in California, 1849-1933* (Berkeley: Institute of Governmental Studies, University of California, 1964), pp. 147-151, 232-240; Donald J. Pisani, *From the Family Farm to Agribusiness: The Irrigation Crusade in California and the West, 1850-1931* (Berkeley: University of California Press, 1984), pp. 275, 283; George H. Hecke, "Ellwood Cooper," *California State Commission of Horticulture Bulletin* 8 (1919): inside title page.

10. Seftel, "Government regulation," pp. 379-392; George Compere, "The origin of plant quarantine in the state of California," typescript, *c.* 1921, 20 pp., HC; Alfred M. Boyce, "Entomology of citrus and its contribution to entomological principles and practices," *Journal of Economic Entomology* 43 (1950): 741-766, on p. 745; Cyril E. Pemberton, "Highlights in the history of entomology in Hawaii, 1778-1963," *Pacific Insects* 6 (1964): 689-729, on pp. 710-711.

11. Leland O. Howard, *A History of Applied Entomology (Somewhat Anecdotal)* (Washington: Smithsonian Miscellaneous Collections, 1930), pp. 121-122, 134-139; Vivian Wiser, "Protecting American agriculture: Inspection and quarantine of imported plants and animals," *U.S. Department of Agriculture Agricultural Economics Report* 266 (1974), pp. 17-19, 28.

12. West's Annotated California Codes: Food and Agricultural Code (St. Paul: West Publishing Co., 1986) §5001; Seftel, "Government regulation," pp. 389-392; John B. Harris to Craw, 18 November 1902, HC; "Neglected orchards may be destroyed under new law," *California Citrograph* 16 (1931): 536.

13. R. P. Cundiff to Craw, 6 June 1903, HC; Henry A. Weinland, *Now the Harvest: Memories of a County Agent* (New York: Exposition Press, 1957), p. 47; West's *California Codes: Food and Agricultural* §5001; Nash, *State Government and Economic Development*, pp. 232-233; Seftel, "Government regulation"; Walter Ebeling, *Subtropical Entomology* (San Francisco: Lithotype Process Co., 1950), pp. 71-72; Coit, *Citrus Fruits*, pp. 107-110.

14. Nash, *State Government and Economic Development*, pp. 239-240; Ray F. Smith, "The origins of integrated control in California, an account of the contributions of Charles W. Woodworth," *Pan-Pacific Entomologist* 50 (1974): 426-440, on pp. 432-433.

15. George Compere, "Origin of fumigation with hydrocyanic acid gas in California," *California Department of Agriculture Bulletin* 11 (1922): 438-442; Charles W. Woodworth, "Orchard fumigation," *California Agricultural Experiment Station Bulletin* 122 (1899), pp. 3-6; Harold Compere, "The red scale and its insect enemies," *Hilgardia* 31 (1961): 173-278, on p. 184.

16. Alfred M. Boyce, "History of the Citrus Research Center and Agricultural Experiment Station," reprinted in idem, *Odyssey of an Entomologist: Adventures on the Farm, at Sea, and in the University* (Riverside, Calif.: UC Riverside Foundation, 1987), pp. 363-378, Lawton and Weathers, "Origins of citrus research," pp. 302-318. For some reason the university campus that grew from the Riverside branch considers 1907 its official starting date, when the Board of Regents approved the lease of additional land for the original site

at the foot of Mt. Rubidoux. See Tom Patterson, *Landmarks of Riverside and the Stories Behind Them* (Riverside, Calif.: Press-Enterprise Co., 1964), p. 179.

17. Henry J. Quayle, "Citrus fruit insects," *California Agricultural Experiment Station Bulletin* 214 (1911): 443–512; idem, "Scale insect parasitism in California," *Journal of Economic Entomology* 4 (1911): 510–515.

18. Harvey A. Lynn, "Founding of the California Fruit Growers Exchange," in Klotz et al., eds., *History of Citrus in the Riverside Area,* pp. 21–24; Ed Ainsworth, *Journey with the Sun: the Story of Citrus in its Western Pilgrimage* (Los Angeles: Sunkist Growers, 1968), pp. 27–40; Jelinek, *Harvest Empire,* pp. 59–60; Starr, *Inventing the Dream,* p. 161; Henry E. Erdman, "The development and significance of California cooperatives," *Agricultural History* 32 (1958): 179–184; Joseph G. Knapp, *The Rise of American Cooperative Enterprise, 1620–1920* (Danville, Ill.: Interstate, 1969), pp. 84–87, 238–247.

19. Morton Rothstein discussed mastery of the market and the advantage of large enterprise for Pacific Coast agriculture in general, in "West coast farmers and the tyranny of distance: Agriculture on the fringes of the world market," *Agricultural History* 49 (1975): 272–280.

20. William W. Cumberland, *Cooperative Marketing: Its Advantages as Exemplified in the California Fruit Growers Exchange* (Princeton: Princeton University Press, 1917), pp. 26–37, 59–69; A. W. McKay et al., "Marketing fruits and vegetables," *Yearbook of the U.S. Department of Agriculture* 1925: 623–710, on pp. 628–635; McWilliams, *Southern California Country,* pp. 211–222 (quotation from p. 218); Belknap, "Era of the lemon," pp. 124–128; Teague, *Fifty Years a Rancher,* pp. 113–114, 141–150 (quotation from p. 141).

21. Belknap, "Era of the lemon," pp. 120–124; Teague, *Fifty Years a Rancher,* pp. 79–81; Cumberland, *Cooperative Marketing,* pp. 55–56; Nash, *State Government and Economic Development,* p. 241; Verne A. Stadtman, ed., *The Centennial Record of the University of California* (Berkeley: University of California Printing Dept., 1967), p. 427.

22. Lynn, "California Fruit Growers Exchange," p. 23; Ainsworth, *Journey with the Sun,* p. 45; Teague, *Fifty Years a Rancher,* pp. 89–90; "Get that beetle" (editorial), *California Citrograph* 22 (1937): 233; Knapp, *Rise of American Cooperative Enterprise,* pp. 333–353; Alan M. Paterson, "Oranges, soot, and science: The development of frost protection in California," *Technology and Culture* 16 (1975): 360–376.

23. Anderson, *Refrigeration in America,* pp. 152–155; Lawton and Weathers, "Origins of citrus research," pp. 307–313; G. Harold Powell, "The decay of oranges while in transit from California," *U.S. Bureau of Plant Industry Bulletin* 123 (1908).

24. Richard J. Orsi, "*The Octopus* reconsidered: The Southern Pacific and agricultural modernization in California, 1865–1915," *California Historical Quarterly* 54 (1975): 197–220, on p. 202; Josephine Kingsbury Jacobs, "Sunkist advertising" (Ph.D. dissertation, University of California, Los Angeles, 1966); Paul S. Armstrong, "Sunkist advertising—how it sells California oranges and lemons," *California Citrograph* 8 (1923): 222–223, 238–289; Harvey A. Levenstein, *Revolution at the Table: The Transformation of the American Diet* (New York: Oxford University Press, 1988), pp. 151–154.

25. Erdman, "California cooperatives," pp. 183–184; Starr, *Inventing the Dream,* pp. 143, 161–163; McKay et al., "Marketing fruits and vegetables," pp. 670–671; J. M. Thompson, "The orange industry—an economic study," *California Agricultural Experiment Station Bulletin* 622 (1938), p. 4.

26. Charles C. Teague, *Ten Talks on Citrus Marketing* (Los Angeles, 1939), pp. 7–12, 21–25; Erdman, "California cooperatives," pp. 180–184; Cumberland, *Cooperative Marketing,* pp. 209–213; John William Lloyd, *Co-operative and Other Organized Methods of Marketing California Horticultural Products* (Urbana: University of Illinois, 1919), pp. 114–115.

27. "Ponder this" (editorial), *California Citrograph* 10 (1924): 41.

28. Advertisement for California Spray-Chemical Company, ibid., 10 (1925): 259.

29. J. A. Prizer, "Pest control and its relation to marketing of citrus," *California Citrograph* 10: (1925): 197, 232; Jacobs, "Sunkist advertising," pp. 27, 31, 337; Cumberland, *Cooperative Marketing*, pp. 105-106. Exhortations to quality were especially common in articles and editorials of the *California Citrograph* in the 1920s.

30. Lloyd, *Co-Operative and Other Organized Methods of Marketing*, pp. 27-28; McKay et al., "Marketing fruits and vegetables," pp. 628, 642-644, 662-689.

31. Raymond L. Spangler, "Standardization and inspection of fresh fruits and vegetables," *U.S. Department of Agriculture Miscellaneous Publication* 604 (1946); Charles C. Teague, "Quality and quantity and their relations to successful marketing of citrus fruits," *California Citrograph* 10 (1925): 194, 214-215; "Supreme Court sustains fruit standardization law," ibid., 16 (1931): 367, 370.

Chapter 3. THE PARASITE CRAZE

1. Riley to Daniel W. Coquillett, 30 March 1891, vol. 57, and 23 June 1891, vol. 58, E5, RG7; Lelong to Rusk, 1 April 1891, AK; Edwin Willits to Lelong, 17 April 1891, AK; Ellwood Cooper, *Bug vs. Bug: Parasitology* (Santa Barbara, Calif., 1913), p. 10; Rusk to Koebele, 29 May 1891, AK.

2. Riley to Koebele, 4 January 1892, AK; Lelong to Koebele, 31 March 1892, AK.

3. Riley to Koebele, 15 June 1892, vol. 67, E5, RG7; Cooper, *Bug vs. Bug*, pp. 4-6; Cooper to Koebele, 24 October 1893, AK; Harry S. Smith, "Biological control of insect pests," in *The Citrus Industry*, vol. 2: *Production of the Crop*, ed. Leon D. Batchelor and Herbert J. Webber (Berkeley: University of California Press, 1948), pp. 597-625, on p. 605.

4. Riley to Koebele, 22 September 1893 and 20 October 1893, AK; Cooper to Koebele, 22 August 1892, AK; Joseph Marsden to Koebele, ? July 1893, AK; Koebele to Howard, 6 June 1894, tray 59, E3, RG7; Howard to Coquillett, 11 August 1892, vol. 70, E5, RG7.

5. Leland O. Howard, *A History of Applied Entomology (Somewhat Anecdotal)* (Washington: Smithsonian Miscellaneous Collections, 1930), pp. 91-93; Riley to Morton, 26 April 1894, box 1, Correspondence 1871-1894, CVR-NAL; "Damaging facts *in re* Board of Horticulture," *Rural Californian* 1895: 144-146; "Prof. C. V. Riley, M.A., Ph.D.," *Entomological News* 6 (1895): 241-243.

6. A. Hunter Dupree, *Science in the Federal Government: A History of Policies and Activities to 1940* (Cambridge, Mass.: Belknap Press of Harvard University Press, 1957), pp. 158-163; Thomas R. Dunlap, *DDT: Scientists, Citizens, and Public Policy* (Princeton: Princeton University Press, 1981), pp. 23-24, 32-37; Leland O. Howard, *The Insect Menace* (New York: Century, 1931).

7. Harold Compere to Edward A. Steinhaus, 11 May 1959, HC.

8. Leland O. Howard, "The parasite element of natural control of injurious insects and its control by man," *Journal of Economic Entomology* 19 (1926): 271-282, on pp. 274-276; idem, *History of Applied Entomology*, pp. 154-155.

9. Harold Compere, "The red scale and its insect enemies," *Hilgardia* 31 (1961): 173-278, on p. 184; Richard L. Doutt, "The historical development of biological control," in *Biological Control of Insect Pests and Weeds*, ed. Paul DeBach (New York: Reinhold, 1964), pp. 21-42, on p. 39.

10. Cooper, *Bug vs. Bug*; George Compere to Cooper, 11 June 1900, HC; G. Compere to Howard, 17 February 1906, HC.

11. Howard to Cooper, 9 March 1895, tray 36, E3, RG7; Craw to Howard, 30

November 1897, and Howard to Craw, 23 November 1897, box 516, SE14, RG7; Howard to Edward Everett Hale, 4 August 1899, tray 115, E6, RG7.

12. H. Compere, "Red scale and its insect enemies," pp. 199–200; idem, manuscripts for "Changing trends and objectives in biological control," 1968, HC; William Wood to G. Compere, 5 January 1899, HC; Entries for 19 April and 14 December 1899, Minutes, Board of Horticulture 1883–1902, CSA.

13. G. Compere to Craw, 17 January 1900, HC; G. Compere to Howard, 30 November 1901, HC.

14. Howard to G. Compere, 12 December 1898, HC.

15. Howard to Paul Marchal, 15 March 1907, tray 132, E6, RG7. Also Howard to Marchal, 26 January 1908, tray 133, E6, RG7.

16. Howard, *History of Applied Entomology,* p. 157; idem, "An apologetic correction," *Science* 73 (1931): 342.

17. G. Compere to Craw, 9 May and 4 July 1900, HC; G. Compere to Craw, 17 January 1901, HC; Craw to Howard, 21 February 1901, box 516, SE14, RG7; Howard to Craw, 2 March 1901, ibid.

18. Craw to G. Compere, 5 March 1901, HC; Alexander Craw, "Horticultural quarantine reports," *Biennial Report of the California State Board of Horticulture* 1901–02: 185–204, on p. 195; Harry S. Smith, "The fundamental importance of life-history data in biological control work," *Journal of Economic Entomology* 19 (1926): 708–714; Paul DeBach and Blair R. Bartlett, "Methods of Colonization, Recovery, and Evaluation," in DeBach, ed., *Biological Control of Insect Pests and Weeds,* pp. 402–426, on p. 411.

19. Leland O. Howard, "Insect parasites of insects," *Proceedings of the Entomological Society of Washington* 26 (1924): 27–46, on p. 28; idem, *History of Applied Entomology,* p. 528; Howard to H. Compere, 10 January 1931, box 56, E35, RG7; Howard, "Apologetic correction."

20. Howard to R. C. L. Perkins, 20 January 1906, tray 40, E6, RG7; Howard to William F. Fiske, 19 October 1918, tray 107, ibid.; Curtis P. Clausen to H. Compere, 19 August 1935, HC; Clausen to Lee A. Strong, 15 November 1933, box 518, SE14, RG7.

21. Leland O. Howard and William F. Fiske, "The importation into the United States of the parasites of the gipsy moth and the brown-tail moth," *U.S. Bureau of Entomology Bulletin* 91 (1911), pp. 36–38; David Rosen and Paul DeBach, "Diaspididae," in *Introduced Parasites and Predators of Arthropod Pests and Weeds: A World Review,* ed. Curtis P. Clausen, Agriculture Handbook no. 480 (Washington: U.S. Department of Agriculture, 1978), pp. 78–128, on p. 124.

22. H. Compere, "Red scale and its insect enemies," p. 202; G. Compere to Cooper, 7 October 1901, HC; Carl Boris Schedvin, *Shaping Science and Industry: A History of Australia's Council for Scientific and Industrial Research, 1926–49* (Sydney: Allen & Unwin, 1987), p. 16; Cooper to George C. Pardee, 28 February and 14 April 1903, GCP; Cooper to G. Compere, 9 February 1904, HC.

23. Howard, *History of Applied Entomology,* p. 372; Howard to Lounsbury, 25 November 1905, tray 129, E6, RG7; Howard to Harry S. Smith, 16 June 1928, box 256, E35, RG7; Harry S. Smith, "What may we expect from biological control?" *Journal of Economic Entomology* 16 (1923): 506–511, on pp. 506–507; George Compere, "A few facts concerning the fruit flies of the world," *California State Commission of Horticulture Bulletin* 1 (1912): 709–730, 842–845, 907–911, 929–932, on pp. 843–845, 907, and typescript, pp. 29–52, HC; Charles P. Lounsbury, "Report of the Government Entomologist for the year 1905," reprint in HC, pp. 98–99; Lounsbury to G. Compere, 14 March 1906, HC.

24. Dupree, *Science in the Federal Government,* pp. 162–163; Howard, *History of Applied Entomology,* p. 160.

25. Howard and Fiske, "Parasites of gipsy moth and brown-tail moth," pp. 47–84; Howard to Fiske, 19 February 1910, E34, RG7.

26. William F. Fiske, "Superparasitism: An important factor in the natural control of insects," *Journal of Economic Entomology* 3 (1910): 88–97; Harry S. Smith, "An attempt to redefine the host relationships exhibited by entomophagous insects," ibid., 9 (1916): 477–486.

27. Thomas R. Dunlap, "The triumph of chemical insecticides in insect control 1890–1920," *Environmental Review* 1, no. 5 (1978): 38–47; Howard to Fiske, 13 December 1910, E34, RG7.

28. Leland O. Howard, "A study in insect parasitism: A consideration of the parasites of the white-marked tussock moth, with an account of their habits and interrelations, and with descriptions of new species," *U.S. Division of Entomology Technical Series* 5 (1897), pp. 6, 48–50.

29. Howard and Fiske, "Parasites of gipsy moth and brown-tail moth," pp. 102–109.

30. William F. Fiske, "The gipsy moth as a forest insect, with suggestions as to its control," *U.S. Bureau of Entomology Circular* 164 (1913); Fiske to Howard, 9 December 1917, E34, RG7.

31. Fiske, "Superparasitism," pp. 88–97.

32. Paul Marchal, "L'équilibre numerique des espèces et ses relations avec les parasites chez les insectes," *Comptes rendus hebdomadaires des séances et memoires de la Société de Biologie* 49 (1897): 129–130. On the logistic equation and its subsequent use in population theory, see Sharon E. Kingsland, *Modeling Nature: Episodes in the History of Population Ecology* (Chicago: University of Chicago Press, 1985), pp. 64–97.

33. Paul Marchal, "The utilization of auxiliary entomophagous insects in the struggle against insects injurious in agriculture," *Popular Science Monthly* 72 (1908): 352–370, 406–419, on pp. 353–358.

34. Charles W. Woodworth, "The theory of the parasitic control of insect pests," *Science* 28 (1908): 227–230. Harold Compere seems to have been the first worker to discover Woodworth's priority for density dependence, but by that time Compere was also a critic. See H. Compere, "Red scale and its insect enemies," pp. 173–278, on p. 184. Woodworth's paper was not generally known until after someone spotted it and sent a copy to another California biological control worker, Carl B. Huffaker, in 1963. Interview with Carl B. Huffaker, 10 June 1987.

35. Boris P. Uvarov, "Insects and climate," *Transactions of the Entomological Society of London* 79 (1931): 1–247, on pp. 156–158; Friedrich Simon Bodenheimer, *Problems of Animal Ecology* (London: Oxford University Press, 1938), p. 81.

36. Joel B. Hagen, "Organism and environment: Frederic Clements's version of a unified physiological ecology," in *The American Development of Biology,* ed. Ronald Rainger, Keith R. Benson, and Jane Maienschein (Philadelphia: University of Pennsylvania Press, 1988), pp. 257–280; idem, *An Entangled Bank: The Origins of Ecosystem Ecology* (New Brunswick: Rutgers University Press, 1992), pp. 20–28; Ronald C. Tobey, *Saving the Prairies: The Life Cycle of the Founding School of American Plant Ecology, 1895–1955* (Berkeley: University of California Press, 1981); Donald Worster, *Nature's Economy: A History of Ecological Ideas* (Cambridge: Cambridge University Press, 1985), pp. 205–253; Victor E. Shelford, *Animal Communities in Temperate America as Illustrated in the Chicago Region,* 2nd ed. (Chicago: University of Chicago Press, 1937), pp. 17–18, 166–168; Robert P. McIntosh, *The Background of Ecology: Concept and Theory* (Cambridge: Cambridge University Press, 1985), pp. 87–88, 146.

37. Howard, "Insect parasites of insects," p. 28; Howard to A. H. Kirkland, 26 January 1907, tray 127, E6, RG7; Cooper to Kirkland, 29 January 1907, tray 387, E31, RG7;

Cooper, *Bug vs. Bug,* pp. 15–21.

38. Craw to James Wilson, 28 November 1904, box 516, SE14, RG7; Howard to John W. Jeffrey, 2 August 1907, tray 124, E6, RG7; Edward O. Essig, "Official entomology in California—Some comments, historical and personal," *California Department of Agriculture Bulletin* 44 (1955): 3–16, on pp. 15–16.

39. Cooper to Craw, 26 August 1903, HC; Cooper to G. Compere, 10 May 1906, HC. Now the park maintenance headquarters, this remains the only building in the park besides the Capitol itself.

40. George E. Mowry, *The California Progressives* (Berkeley: University of California Press, 1951); Spencer C. Olin, Jr., *California's Prodigal Sons: Hiram Johnson and the Progressives, 1911–1917* (Berkeley: University of California Press, 1968); H. Compere, "Red scale and its insect enemies," p. 186; Cooper to G. Compere, 4 June 1907, HC; Howard to Jeffrey, 24 September 1907, tray 124, E6, RG7; Marlatt to Jeffrey, 7 July 1908, ibid.; Jeffrey to Howard, 15 November 1907 and 15 June 1908, box 63, E2, RG7.

41. Howard, *History of Applied Entomology,* p. 155; Howard to Jeffrey, 22 June and 5 August 1908, tray 124, E6, RG7; Jeffrey to Howard, 30 June 1908, box 63, E2, RG7; Howard to Philip H. Timberlake, 29 April 1911, PHT; Carnes to G. Compere, 20 September 1908, HC.

42. Carnes to C. K. McClatchy, 3 March 1909, HC; Stanley E. Flanders, "George Compere, pioneer in the biological control of red scale," *California Citrograph* 34 (1949): 160–162; Howard to Jeffrey, 25 October 1909, E34, RG7; Leland O. Howard, "Two new aphelinine parasites of scale insects," *Entomological News* 21 (1910): 162–163; Howard to Carnes, 1 April 1910, box 515, SE14, RG7; Carnes to Howard, 14 March 1910, ibid.

43. Howard, "Parasite element of natural control," p. 276; Flanders, "George Compere"; H. Compere, "Red scale and its insect enemies," pp. 204–207; G. Compere to Albert J. Cook, 29 August 1912, HC.

44. G. Compere to Jacob Kotinsky, 4 June 1909, HC; G. Compere to Frederick Maskew, 10 June 1912, HC; Charles C. Teague et al., "Report of the conditions of the California State Commission of Horticulture," typescript, 13 August 1913, pp. 35–36, filed under Cook, A. J., EOE; "Twenty-first Mediterranean fruit fly seminar," 1 April 1931, pp. 2–3, box 1075, SE62, RG7; "History and distribution of the Mediterranean fruit fly," in "Seventy-sixth Mediterranean fruit fly seminar," 3 August 1932, pp. 3–4, box 1076, ibid.; Henry H. P. Severin, "The introduction, methods of control, spread, and migration of the Mediterranean fruit fly in the Hawaiian Islands," *California State Commission of Horticulture Bulletin* 1 (1912): 558–565.

45. Edward K. Carnes, "Insectary division. Report for the month of May," *California State Commission of Horticulture Bulletin* 1 (1912): 395–400, on p. 395; idem, "Collecting ladybirds (Coccinellidae) by the ton"; ibid., pp. 71–81; *California State Commission of Horticulture Bulletin* 2 (1913): 626.

46. Charles Elton, *The Ecology of Invasions by Animals and Plants* (London: Methuen, 1958), pp. 83–84; Edward D. Beechert, *Working in Hawaii: A Labor History* (Honolulu: University of Hawaii Press, 1985), pp. 133–139, 178–180; J. H. Galloway, *The Sugar Cane Industry: An Historical Geography from its Origins to 1914* (Cambridge: Cambridge University Press, 1989), pp. 219–228; Thomas J. Osborne, *"Empire Can Wait": American Opposition to Hawaiian Annexation, 1893–1898* (Kent, Ohio: Kent State University Press, 1981), pp. 17–27; Walter M. Giffard, "A review of the organization of the Hawaiian Entomological Society and brief mention of some of the more notable achievements in Hawaii by its members," *Proceedings of the Hawaiian Entomological Society* 4 (1920): 363–373; Leland O. Howard, "On the Hawaiian work in introducing beneficial insects," *Journal of Economic Entomology* 9 (1916): 172–179.

47. Richard A. Overfield, "The agricultural experiment station and Americanization: The Hawaiian experience, 1900–1910," *Agricultural History* 60, no. 2 (1986): 256–266.

48. Cyril E. Pemberton, "History of the Entomology Department Experiment Station, HSPA 1904–45," *Hawaiian Planters Record* 52, no. 1 (1948): 53–90, on pp. 56–67; idem, "The biological control of insects in Hawaii," *Proceedings of the Seventh Pacific Science Congress* 4 (1953): 220–223.

49. Overfield, "Agricultural experiment station and Americanization," p. 265; Beechert, *Working in Hawaii*, p. 182; Pemberton, "Biological control of insects in Hawaii," pp. 220–221.

50. Jack R. Coulson, "Recent history: 1961–77," in "The gypsy moth: research toward integrated pest management," ed. Charles C. Doane and Michael L. McManus, *U.S. Department of Agriculture Technical Bulletin* 1584 (1981): 302–310.

51. Jerry Woods Weeks, "Florida gold: The emergence of the Florida citrus industry, 1865–1895" (Ph.D. dissertation, University of North Carolina, Chapel Hill); Joseph G. Knapp, *The Advance of American Cooperative Enterprise: 1920–1945* (Danville, Ill.: Interstate, 1973), pp. 419–420; idem, *The Rise of American Cooperative Enterprise, 1620–1920* (Danville, Ill.: Interstate, 1969), p. 81; Howard Seftel, "Government regulation and the rise of the California fruit industry: The entrepreneurial attack on fruit pests, 1880–1920," *Business History Review* 59 (1985): 369–402; James T. Hopkins, *Fifty Years of Citrus: The Florida Citrus Exchange, 1909–1959* (Gainesville: University of Florida Press, 1960), pp. 146–148, 179–180.

52. Austin W. Morrill and Ernest A. Back, "Natural control of white flies in Florida," *U.S. Bureau of Entomology Bulletin* 102 (1912); Joseph R. Watson, "Citrus insects and their control," *Florida Agricultural Experiment Station Bulletin* 183 (1926): 289–423, on pp. 340–366; Russell S. Woglum, "Report of a trip to India and the Orient in search of the natural enemies of the citrus white fly," *U.S. Bureau of Entomology Bulletin* 120 (1913): 15–40; Reece I. Sailer et al., "Dissemination of the citrus whitefly (Homoptera: Aleyrodidae) parasite *Encarsia lahorensis* (Howard) (Hymenoptera: Aphelinidae) and its effectiveness as a control agent in Florida," *Bulletin of the Entomological Society of America* 30 (1984): 36–39.

53. Howard to Smith, 20 June 1921, HC; Howard, "Parasite element of natural control," p. 276.

Chapter 4. BIOLOGICAL CONTROL COMES OF AGE

1. Edwin T. Layton, Jr., "Mirror-image twins: The communities of science and technology in 19th-century America," *Technology and Culture* 12 (1971): 562–580; idem, "Technology as knowledge," ibid., 15 (1974): 31–41; idem, "Through the looking glass, or news from Lake Mirror Image," ibid., 28 (1987): 594–607; Barry Barnes, "The science-technology relationship: A model and a query," *Social Studies of Science* 12 (1982): 166–172; John M. Staudenmaier, S.J., *Technology's Storytellers: Reweaving the Human Fabric* (Cambridge, Mass.: Society for the History of Technology and MIT Press, 1985), pp. 85–103; Judith P. Swazey and Karen Reeds, *Today's Medicine, Tomorrow's Science: Essays on Paths of Discovery in the Biomedical Sciences* (Washington: Department of Health, Education, and Welfare, Public Health Service, National Institutes of Health, 1978).

2. Deborah Fitzgerald, *The Business of Breeding: Hybrid Corn in Illinois, 1890–1940* (Ithaca, N.Y.: Cornell University Press, 1990), p. 42.

3. Edward O. Essig to Albert J. Cook, 10 December 1910, EOE; Edwin T. Earl to Johnson, 3 May 1911, HWJ; Harold Compere, "The red scale and its insect enemies," *Hilgardia* 31 (1961): 173–278, on p. 207.

4. Arnold Mallis, *American Entomologists* (New Brunswick: Rutgers University Press, 1971), pp. 138–141; Harry W. Lawton and Lewis G. Weathers, "The origins of citrus research in California," in *The Citrus Industry*, vol. 5: *Crop Protection, Postharvest Technology, and Early History of Citrus Research in California*, ed. Walter Reuther, E. Clair Calavan, and Glenn E. Carman (Berkeley: University of California Division of Agriculture and Natural Resources, 1989), pp. 281–335, on p. 304; Charles C. Teague, "A tribute to the memory of Dr. A. J. Cook," *California Citrograph* 6 (1921): 282, 293; Cook to Benjamin Ide Wheeler, 16 September 1900, BIW.

5. Essig to Cook, 16 April 1911, EOE; Charles C. Teague et al., "Report of the conditions of the California State Commission of Horticulture," typescript, 13 August 1913, filed under Cook, A. J., EOE; Jeffrey to Chester H. Rowell, no date [1913], CHR; Johnson to Rowell, 26 June 1913, CHR.

6. Howard to Cook, 7 December 1905, tray 98, E6, RG7; Cook to Howard, 14 February 1906, tray 390, E31, RG7; Leland O. Howard, *A History of Applied Entomology (Somewhat Anecdotal)*, Smithsonian Miscellaneous Collections, no. 84 (Washington: Smithsonian Institution, 1930), p. 21.

7. Albert J. Cook, "The State Horticultural Commission," *California State Commission of Horticulture Bulletin* 1 (1911): 30–34; Compere, "Red scale and its insect enemies," p. 207.

8. G. Compere to Cook, 29 August and 18 January 1912, HC; Essig to G. Compere, 10 September 1912, EOE; Teague et al., "Conditions of State Commission of Horticulture"; G. Compere to Carnes, 23 December 1912, HC; Harold Compere to Harry S. Smith, 21 March 1917, HC.

9. Mallis, *American Entomologists*, pp. 191–195, 501; Howard, *History of Applied Entomology*, p. 103; Myron H. Swenk, "In memoriam—Lawrence Bruner," *Nebraska Bird Review* 5 (1937): 35–48; interview with Lawrence F. Smith, 27 May 1987; interview with Richard A. Smith, 30 May 1987.

10. Bruner to Howard, 22 July 1901, box 7, E2, RG7.

11. Fiske to Howard, 20 March 1910, E34, RG7.

12. Harry S. Smith, "The chalcidoid genus *Perilampus* and its relation to the problem of parasite introduction," *U.S. Bureau of Entomology Technical Series* 19 (1912): 33–69; idem, "The habit of leaf oviposition among the parasitic Hymenoptera," *Psyche* 24 (1917): 63–68.

13. Smith to Bruner, 7 March 1910, HSS; Smith to W. I. Farley, 14 July 1908, HSS; Fiske to Howard, 3 June 1910, E34, RG7; Richard Smith, interview.

14. Harry S. Smith, "The alfalfa weevil," *California State Commission of Horticulture Bulletin* 6 (1917): 295–297; Thomas R. Chamberlin, "The introduction and distribution of the alfalfa weevil parasite, *Bathyplectes curculionis* (Thoms.), in the United States," *Journal of Economic Entomology* 19 (1926): 302–310; Abraham E. Michelbacher, "The present status of the alfalfa weevil in California," *California Agricultural Experiment Station Bulletin* 677 (1943), pp. 19–21; Smith to Charles H. T. Townsend, 15 March 1913, HSS.

15. "Harry S. Smith retires," *California Citrograph* 36 (1951): 380–382; Cook to Howard, 9 November 1912, E34, RG7; Howard to Cook, 2 December 1912, ibid.; Howard to Smith, 2 December 1912, HC; Francis M. Webster to Cook, 11 December 1912, E34, RG7; Smith to Howard, 13 December 1912, HSH.

16. Howard to Cook, 21 January 1913, E34, RG7; Howard to Smith, 11 February 1913, HC; Smith to Howard, 9 March and 16 June 1914, HC.

17. Smith to Howard, 29 October 1915, E34, RG7; Harry S. Smith, "What may we expect from biological control?" *Journal of Economic Entomology* 16 (1923): 506–511.

18. Howard, *History of Applied Entomology*, p. 155; Smith to Howard, 16 June 1914,

HC; Smith to Howard, 6 January 1920, E34, RG7.

19. Harry S. Smith and Everett J. Vosler, "*Calliephialtes* in California," *California State Commission of Horticulture Bulletin* 3 (1914): 195-211; Harry S. Smith, "Some misconceptions," ibid., 4 (1915): 269-270; Smith to Earle Mills, 25 April 1914, HSS.

20. Harry S. Smith, "Biennial report of the Insectary Division, State Commission of Horticulture, 1917-18," *California State Commission of Horticulture Bulletin* 8 (1919): 44-51, on p. 50; E. J. Branigan, "Vedalia vs. Icerya on pears," ibid., 4 (1915): 107-108.

21. Harry S. Smith, "The season's work with *Hippodamia convergens,*" *California State Commission of Horticulture Bulletin* 3 (1914): 77-78; idem, "Report of the State Insectary," ibid., 4 (1915): 542-543; Smith to Charles W. Woodworth, 15 June 1915, HSS; Smith to Howard, 14 April 1917, HC; Smith to H. Compere, 12 May 1917, HC.

22. Howard to Philip H. Timberlake, 29 April 1911, PHT; W. M. Davidson, "The convergent ladybird beetle (*Hippodamia convergens* Guerin) and the barley-corn aphis (*Aphis maidis* Fitch)," *California State Commission of Horticulture Bulletin* 8 (1919): 23-26; idem, "Observations and experiments on the dispersion of the convergent lady-beetle (*Hippodamia convergens* Guérin) in California," *Transactions of the American Entomological Society* 50 (1925): 163-175; Melville H. Hatch and Cherie Tanasse, "The liberation of *Hippodamia convergens* in the Yakima valley of Washington, 1943 to 1946," *Journal of Economic Entomology* 41 (1948): 993; Paul DeBach and Kenneth S. Hagen, "Manipulation of entomophagous species," in *Biological Control of Insect Pests and Weeds,* ed. Paul DeBach (New York: Reinhold, 1964), pp. 429-458, on pp. 443-444.

23. Entry for 1 June 1913, Minutes of Cabinet Meetings, California State Commission of Horticulture, CSA; Harry S. Smith, "Mealy bug parasites in the Far East," *California State Commission of Horticulture Bulletin* 3 (1914): 26-29.

24. Harry S. Smith, "On the life-history and successful introduction of the Sicilian mealy-bug parasite," *Journal of Economic Entomology* 10 (1917): 262-268; Smith to Frederick Maskew, 24 December 1914, HSS.

25. E. J. Branigan, "A satisfactory method of rearing mealybugs for use in parasite work," *California State Commission of Horticulture Bulletin* 5 (1916): 304-306; Harry S. Smith, "A sublaboratory of the Insectary in the south," ibid., 307; H. Morton Armitage, "Controlling mealybugs by the use of their natural enemies," ibid., 8 (1919): 257-260; Byron M. Lelong, *Culture of the Citrus in California* (Sacramento: A. G. Johnston, 1902), p. 262.

26. G. DeLotto, "The Pseudococcidae (Hom.: Coccoidea) described by C. K. Brain from South Africa," *Bulletin of the British Museum (Natural History), Entomology* 7 (1958): 79-120, on p. 96; Howard L. McKenzie, "Fourth taxonomic study of California mealybugs, with additional species from North America, South America, and Japan (Homoptera: Coccoidea: Pseudococcidae)," *Hilgardia* 35 (1964): 211-272, on p. 255, n. 9; Curtis P. Clausen, "Mealy bugs of citrus trees," *California Agricultural Experiment Station Bulletin* 123 (1915): 17-48, on p. 30; Gordon F. Ferris, "Observations on some mealy-bugs (Hemiptera; Coccidae)," *Journal of Economic Entomology* 12 (1919): 292-299.

27. Harry S. Smith and H. Morton Armitage, "Biological control of mealybugs in California," *California Department of Agriculture Bulletin* 9 (1920): 104-158; idem, "The biological control of mealybugs attacking citrus," *California Agricultural Experiment Station Bulletin* 509 (1931); H. Morton Armitage, "Report of the biological control work directed against the mealybugs," *California Department of Agriculture Bulletin* 9 (1920): 441-451; idem, "The citrophilus mealybug, *Pseudococcus gahani* Green, as a major pest of citrus in southern California, *Journal of Economic Entomology* 17 (1924): 554-561; idem, "Timing field liberations of Cryptolaemus in the control of citrophilus mealybug in the infested citrus orchards of southern California," ibid., 22 (1929): 910-915; Harry S. Smith, "The

commercial development of biological control in California," ibid., 18 (1925): 147–152; Anson A. Brock, "A brief summary of what has been done by the Orange County Pest Control Association," typescript, 17 October 1924, HC.

28. Smith and Armitage, "Biological control of mealybugs attacking citrus," p. 70; Armitage to Howard, 16 June 1924, E34, RG7; "H. M. Armitage," *California Department of Agriculture Bulletin* 20 (1931): 598.

29. Harry S. Smith, untitled manuscript on grape mealybug, filed under H. S. Smith, UCR-BC; Smith and Armitage, "Biological control of mealybugs attacking citrus," p. 72; Blair R. Bartlett, "Pseudococcidae," in *Introduced Parasites and Predators of Arthropod Pests and Weeds: A World Review,* ed. Curtis P. Clausen, Agriculture Handbook no. 480 (Washington: U. S. Department of Agriculture, 1978), pp. 137–170, on p. 167; Smith to L. R. Boedefeld, 7 January 1914, HSS; Smith to Howard, 1 December 1914, HC.

30. Smith to Howard, 24 November 1919, E34, RG7.

31. Harry S. Smith, "Beet leaf-hopper parasites," *California State Commission of Horticulture Bulletin* 5 (1916): 299; idem, "An Australian expedition from the Insectary," ibid., 6 (1917): 33; Alfred S. Eichner, *The Emergence of Oligopoly: Sugar Refining as a Case Study* (Baltimore: Johns Hopkins Press, 1969), pp. 228–263; Everett J. Vosler, "Some work of the Insectary Division in connection with the attempted introduction of natural enemies of the beet leafhopper," ibid., 8 (1919): 231–239; Henry H. P. Severin, "Natural enemies of beet leafhopper (*Eutettix tenella* Baker)," *Journal of Economic Entomology* 17 (1924): 369–377; Harold Compere, "History of the introduction of *Quaylea whittieri* in California from Australia," typescript, HC; Harry S. Smith, "Everett Jay Vosler," *California State Commission of Horticulture Bulletin* 7 (October 1918), inside title page.

32. Leland O. Howard and William F. Fiske, "The importation into the United States of the parasites of the gypsy moth and the brown-tail moth," *U.S. Bureau of Entomology Bulletin* 91 (1911), p. 152; Harry S. Smith, "Biological control of the black scale in California," *California Department of Agriculture Bulletin* 10 (1921): 127–137; Harry S. Smith and Harold Compere, "The life-history and successful introduction into California of the black scale parasite, *Aphycus lounsburyi* How.," *California Department of Agriculture Bulletin* 9 (1920): 310–320.

33. Smith to Howard, 3 February and 12 July 1921, E34, RG7; Armitage to Smith, 12 October 1921, ibid.; Armitage to Smith, 18 June 1921, HC; Armitage to Smith, 17 July 1920, Records of the Plant Pest Control Service, F3742: 2294, CSA; Smith to H. Compere, 12 November 1920, ibid., F3742: 2295; Smith to H. Compere, 14 June 1921, HC; Harold Compere, "The black scale problem," *California Cultivator* 59 (1922): 29–30; Compere's notes, dated 1972, on a copy of this article, UCR-BC.

34. Smith to Howard, 27 July 1921, E34, RG7; Harry S. Smith, "Biological control work," *California Department of Agriculture Bulletin* 12 (1923): 334–342.

35. Smith to Howard, 18 March 1919, E34, RG7; Howard to Smith, 24 March 1919, Records of the Plant Pest Control Service, F3742: 2311, CSA.

36. Lounsbury to Henry J. Quayle, 16 November 1910, box 2, HJQ; Quayle to Lounsbury, 30 September 1910, ibid.

37. Blair R. Bartlett, "Coccidae," in Clausen, ed., *Introduced Parasites and Predators,* pp. 57–74, on pp. 68–70; Compere, "Red scale and its insect enemies," pp. 210–211; Edward O. Essig, *A History of Entomology* (New York: Macmillan, 1931), pp. 381–382; Smith to Howard, 15 June 1921, HC; Rust to Smith, 2 May and 4 July 1927, HSS; Smith to Howard, 11 June 1928, box 256, E35, RG7.

38. Charles W. Woodworth, "Insect pests in California," *Journal of the Department of Agriculture of Western Australia* 8 (1903): 564–567; Compere, "Red scale and its insect enemies," p. 184; Ray F. Smith, "The origins of integrated control in California, an account

of the contributions of Charles W. Woodworth," *Pan-Pacific Entomologist* 50 (1974): 426–440; John H. Perkins, *Insects, Experts, and the Insecticide Crisis: The Quest for New Pest Management Strategies* (New York: Plenum, 1981), pp. 73–75; Thomas R. Dunlap, *DDT: Scientists, Citizens, and Public Policy* (Princeton: Princeton University Press, 1981), pp. 35–36; Henry J. Quayle, "A statistical study of brown scale parasitism," *Science* 27 (1908): 788–789; idem, "*Scutellista cyanea* Motsch.," *Journal of Economic Entomology* 3 (1910): 446–451.

39. Harry S. Smith, "Report of the Bureau of Pest Control," *California Department of Agriculture Bulletin* (1921): 570–597, on p. 570; Smith to Howard, 13 December 1915 and 24 November 1920, E34, RG7; Howard to Smith, 6 March 1916, HC; Smith to H. Compere, 14 January 1921, HC; Smith to Howard, 28 March 1921, E34, RG7.

40. Smith to Ernest R. Barber, 21 March 1923, HSS; "Harry S. Smith retires," p. 380; Smith to Howard, 10 November 1922, E34, RG7; Smith to Clarence M. Haring, 6 February 1923, HSS; Essig to Thomas F. Hunt, no date, Essig, E. O.—Entomology and Parasitology, UCB, 14: 13, UCA-CA; H. Compere to Smith, 9 April 1923, HC; Smith to Edward P. Van Duzee, 23 May 1923, HSS; Haring to Teague, 8 May 1923, HSS.

41. Smith to Haring, 29 May 1923, HSS; Smith to Howard, 12 June 1923, HSS; Smith to J. T. Barrett, 18 April 1925, 1924: 35, UCA-PF; Elmer D. Merrill to Smith, 6 April 1927, HSS; Smith to Merrill, 11 April 1927, HSS.

42. Smith to Timberlake, 26 December 1923, PHT; Timberlake to Smith, 7 January 1924, PHT.

43. Howard to Timberlake, 11 December 1925, PHT; Howard to Smith, 23 December 1926, HC; Timberlake to Leon D. Batchelor, 21 June 1930, Philip H. Timberlake file, UCR-DE; interview with Richard L. Doutt, 30 May 1987; Alfred M. Boyce, *Odyssey of an Entomologist: Adventures on the Farm, at Sea, and in the University* (Riverside, Calif.: UC Riverside Foundation, 1987), p. 253; Smith to Carl B. Huffaker, 6 February 1950, EAS.

44. W. A. Baker, W. G. Bradley, and Charles A. Clark, "Biological control of the European corn borer in the United States," *U. S. Department of Agriculture Technical Bulletin* 983 (1949); Walter E. Fleming, "Biological control of the Japanese beetle," *U. S. Department of Agriculture Technical Bulletin* 1383 (1968); Clausen to Lee A. Strong, 15 November 1933, box 518, SE14, RG7; Leland O. Howard, "Insect parasites of insects," *Proceedings of the Entomological Society of Washington* 26 (1924): 27–46; on pp. 43–44; Curtis P. Clausen, "Biological control of insect pests in the continental United States," *U.S. Department of Agriculture Technical Bulletin* 1139 (1956), pp. 23–25, 42–43.

45. Curtis P. Clausen and Paul A. Berry, "The citrus blackfly in Asia, and the importation of its natural enemies into tropical America," *U. S. Department of Agriculture Technical Bulletin* 320 (1932): 1–22; Lee A. Strong, "Establishment of a Division of Foreign Parasite Introduction," 5 May 1934, box 518, SE14, RG7.

46. For more detail on USDA efforts in biological control during this period, see Richard C. Sawyer, "Monopolizing the insect trade: Biological control in the USDA, 1888–1951," *Agricultural History* 64, no. 2 (1990): 271–285; Reece I. Sailer, "A look at the USDA's biological control of insect pests: 1888 to present," *Agricultural Science Review* 10, no. 4. (1972): 15–27.

47. Harold Compere, "An appraisal of Silvestri's work in the Orient for the University of California, some misidentifications corrected, and two forms of *Casca* described as new species," *Bollettino del laboratorio di zoologia generale e agraria della facolta agraria in Portici* 33 (1953): 35–46; Harry S. Smith, "Status of biological control work of citrus pests," *California Citrograph* 11 (1926): 414, 426–431.

48. Harry S. Smith and Harold Compere, "The introduction of new insect enemies of the citrophilus mealybug from Australia," *Journal of Economic Entomology* 21 (1928): 664–669.

49. "Agricultural patrols by airplane," *California Department of Agriculture Bulletin* 8 (1919): 101–103. After using L. O. Howard's endorsement to get the bollworm survey program approved, Compere made the first flight just a few days before he was discharged in January 1919. His roommate, William Tillisch, took over, and was killed on a scouting flight in August. See H. Compere to Howard, 29 November 1918, HC; Howard to Division of Military Aeronautics, War Department, 9 December 1918, HC; Compere to Smith, 15 January 1919; note on Tillisch's death, HC.

50. Harold Compere, "Successful importation of five new natural enemies of citrophilus mealybug from Australia," *California Citrograph* 13 (1928): 318, 346–349; Harold Compere and Harry S. Smith, "The control of the citrophilus mealybug, *Pseudococcus gahani*, by Australian parasites," *Hilgardia* 6 (1932): 585–618.

51. Gordon F. Ferris, "Mealybugs," *California Department of Agriculture Bulletin* 16 (1927): 336–342; Smith to Ferris, 3 November 1927, HSS; Ferris to Smith, 7 November 1927, HSS; Smith to Compere, 29 December 1927, HC; Compere to Ferris, 24 May 1928, HC.

52. Compere and Smith, "Control of citrophilus mealybug"; Harry S. Smith, "Is citrophilus mealybug under permanent control?" *California Citrograph* 16 (1931): 429, 448; Harry S. Smith and Harold Compere, "Introduced parasites successfully control the citrophilus mealybug," *Journal of Economic Entomology* 24 (1931): 942–945; Smith to Compere, 23 April 1930, HC; Compere to Stanley E. Flanders, 25 May 1931, HC; Bartlett, "Pseudococcidae," p. 162.

53. Paolo S. A. Palladino, "Entomology and ecology: The ecology of entomology" (Ph.D. dissertation, University of Minnesota, 1989), pp. 100–124, quotation from p. 103.

54. Harry S. Smith, "On some phases of preventive entomology," *Scientific Monthly* 29 (1929): 177–184; idem, "What may we expect from biological control?"; idem, "Principles of biological control," typescript, Harry S. Smith Miscellaneous file, UCR-BC. This typescript is not dated, but Smith must have written it before he reached the multiple parasitism conclusions he published in 1929.

55. Sharon E. Kingsland, *Modeling Nature: Episodes in the History of Population Ecology* (Chicago: University of Chicago Press, 1985), p. 100; Smith to Webster, 16 November 1912, PHT.

56. Kingsland, *Modeling Nature*, pp. 100–102; W. Robin Thompson, "La théorie mathématique de l'action des parasites entomophages," *Revue générales des sciences pures et appliquées* 34 (1923): 202–210; idem, "A criticism of the 'sequence' theory of parasitic control," *Annals of the Entomological Society of America* 16 (1923): 115–128; Howard, "Insect parasites of insects."

57. Kingsland, *Modeling Nature*, pp. 102–127.

58. Smith to Royal N. Chapman, 30 April 1929 (information courtesy Sharon E. Kingsland); Smith to Thompson, 23 January 1939, CAF.

59. Smith, "Principles of biological control."

60. Kingsland, *Modeling Nature*, pp. 95, 128–129; Royal N. Chapman, "The potentialities of entomology," *Science* 69 (1929): 413–418; idem, *Animal Ecology with Especial Reference to Insects* (New York: McGraw-Hill, 1931), p. 183–195.

61. Harry S. Smith, "The role of biotic factors in the determination of population densities," *Journal of Economic Entomology* 28 (1935): 873–898, on pp. 878–880.

62. Cyril E. Pemberton and Harold F. Willard, "Interrelations of fruit-fly parasites in Hawaii," *Journal of Agricultural Research* 12 (1918): 285–296.

63. Harry S. Smith, "Multiple parasitism: Its relation to the biological control of insect pests," *Bulletin of Entomological Research* 20 (1929): 141–149; Cyril E. Pemberton and Harold F. Willard, "Parasitism of the larvae of the Mediterranean fruit fly in Hawaii during 1916," *Report of the Board of Commissioners of Agriculture and Forestry of the Territory*

of Hawaii 1915-16: 111-118.

64. Smith to H. Compere, 25 October 1927, HSS.

65. Harry S. Smith, "The utilization of entomophagous insects in the control of citrus pests," *Transactions of the Fourth International Congress of Entomology* 2 (1929): 191-198; William H. Thorpe, "The biology, post-embryonic development, and economic importance of *Cryptochaetum iceryae* (Diptera, Agromyzidae) parasitic on *Icerya purchasi* (Coccidae, Monophlebini)," *Proceedings of the Zoological Society of London* 1930: 929-971; Compere and Smith, "Control of citrophilus mealybug," pp. 612-613; Curtis P. Clausen, "Insect parasitism and biological control," *Annals of the Entomological Society of America* 29 (1936): 201-223, on pp. 204-205; W. Robin Thompson, "The principles of biological control," *Annals of Applied Biology* 17 (1930): 306-338, on pp. 331-332. In Italy the issue of multiple introduction split economic entomology into two factions. Filippo Silvestri, who collected medfly parasites for Hawaii, advocated the use of all possible parasites, while his mentor, Antonio Berlese, bitterly opposed him and took the position of Pemberton and Willard. See Howard, *History of Applied Entomology,* pp. 254-255.

66. Thompson, "Criticism of 'sequence theory.'"

67. Boris P. Uvarov, "A revision of the genus *Locusta,* L. (=*Pachytylus,* Fieb.), with a new theory as to the periodicity and migrations of locusts," *Bulletin of Entomological Research* 12 (1921): 135-163; idem, "Insects and climate," *Transactions of the Entomological Society of London* 79 (1931): 1-247 (quotation from p. 158).

68. Friedrich Simon Bodenheimer, "Welche Faktoren regulieren die Individuenzahl einer Insektenert in der Natur?" *Biologisches Zentralblatt* 48 (1928): 714-739; idem, "Über die Grundlagen einer allgemeinen Epidemiologie der Insektenkalamitäten," *Zeitschrift für Angewandte Entomologie* 16 (1930): 433-450; idem, *Problems of Animal Ecology* (London: Oxford University Press, 1938), p. 81.

69. Friedrich Simon Bodenheimer, *A Biologist in Israel* (Jerusalem: Turin Press, 1959), pp. 103-104; Compere, "Red scale and its insect enemies," pp. 258-259; Harry S. Smith and Paul DeBach, "The measurement of the effect of entomophagous insects on population densities of their hosts," *Journal of Economic Entomology* 35 (1942): 845-849.

70. M. E. Solomon, "Dynamics of insect populations," *Annual Review of Entomology* 2 (1957): 121-142, on p. 139; Carl B. Huffaker, "The concept of balance in nature," *Proceedings of the Tenth International Congress of Entomology* 2 (1958): 625-636, on p. 635; Kingsland, *Modeling Nature,* p. 173.

71. Smith to Alexander John Nicholson, 13 July 1933, AJN.

72. Smith, "Role of biotic factors," pp. 873-874; interview with Alfred M. Boyce, 13-14 May 1987; interview with Paul DeBach, 9 July 1987; Kingsland, *Modeling Nature,* pp. 81-82.

73. Smith, "Role of biotic factors," pp. 880-890; Vito Volterra, "Variations and fluctuations in the number of individuals in animal species living together," trans. Mary Evelyn Wells, in Chapman, *Animal Ecology,* pp. 409-448, on p. 415; Howard and Fiske, "Parasites of gipsy moth and brown-tail moth," p. 107; W. Robin Thompson, "A contribution to the study of biological control and parasitic introduction in continental areas," *Parasitology* 20 (1928): 90-112; idem, *The Biological Control of Insect and Plant Pests: A Report on the Organization and Progress of the Work of Farnham House Laboratory* (London: King's Printer, 1930), pp. 58-64.

74. Smith, "Role of biotic factors," pp. 875-879, 890-895.

75. Ibid., pp. 877, 895-896; Harry S. Smith and Stanley E. Flanders, "Is Trichogramma becoming a fad?" *Journal of Economic Entomology* 24 (1931): 666-672; Smith, "Multiple parasitism."

76. Smith, "Multiple parasitism," pp. 142-143; idem, "The influence of civilization on

the insect fauna by purposeful introductions," *Annals of the Entomological Society of America* 26 (1933): 518–528, on p. 522; idem, "Role of biotic factors," pp. 893–894.

77. Interview with Harvey L. Sweetman, 20 June 1988; Harvey L. Sweetman, *The Biological Control of Insects* (Ithaca: Comstock, 1936), pp. 226–229, 277–278.

78. W. Robin Thompson, "On the relative value of parasites and predators in the biological control of insect pests," *Bulletin of Entomological Research* 19 (1929): 343–350.

79. Harry S. Smith, review of Sweetman, *The Biological Control of Insects, Journal of Economic Entomology* 30 (1937): 218–220. The textbook was *so* faithful to sources such as Thompson that Smith politely accused Sweetman of plagiarism.

80. The following all expressed this sentiment in interviews: DeBach; Carl B. Huffaker (10 June 1987); Doutt; Theodore W. Fisher (28 May 1987); Kenneth S. Hagen (19 June 1987); Irvin M. Hall (30 July 1987); Charles E. Kennett (17 June 1987).

81. Smith, "Multiple parasitism," p. 148; idem, review of Sweetman, quotation from p. 219.

82. Smith, "Role of biotic factors," pp. 878–880; idem, "Influence of civilization," pp. 522–523; idem, "Ecological aspects of insect population dynamics," unpublished paper read at the 1955 meeting of the Entomological Society of America, typescript in possession of Kenneth S. Hagen, p. 23.

83. William C. Kimler, "Advantage, adaptiveness, and evolutionary ecology," *Journal of the History of Biology* 19 (1986): 215–233; Stanley E. Flanders, "The organization of biological control and its historical development," *Mededelingen van de Landbouwhogeschool en de Opzoekingsstations van den Staat te Gent* 20 (1955): 257–270, on p. 270; A. John Nicholson, "A new theory of mimicry in insects," *Australian Zoologist* 5 (1927): 10–104, on pp. 81–87; Kingsland, *Modeling Nature*, pp. 116–117.

84. Kingsland, *Modeling Nature*, pp. 117–122; A. John Nicholson, "The balance of animal populations," *Journal of Animal Ecology* 2 (1933): 132–178.

85. Harry S. Smith, "Insect populations in relation to biological control," *Ecological Monographs* 9 (1939): 311–320; on pp. 318–320; Nicholson to H. Compere, 14 September 1933, HC; Nicholson to Frank N. Egerton, 25 May 1961, AJN; Nicholson, "Balance of animal populations," p. 163.

86. Bailey to Nicholson, 30 July 1937, AJN.

87. Kingsland, *Modeling Nature*, pp. 7, 134–142; Thompson to Nicholson, 23 October 1933, AJN; W. Robin Thompson, "Biological control and the theories of the interactions of populations," *Parasitology* 31 (1939): 299–388; Thompson to Smith, 23 December 1938, CAF.

88. W. Robin Thompson, "On natural control," *Parasitology* 21 (1929): 269–281; idem, "Biological control and theories of interactions."

89. Idem, "Can economic entomology be an exact science?" *Canadian Entomologist* 80 (1948): 49–55; idem, "The fundamental theory of natural and biological control," *Annual Review of Entomology* 1 (1956): 379–402; A. John Nicholson, "Fluctuation of animal populations," *Report of the Australia and New Zealand Association for the Advancement of Science* 26 (1947): 134–147; idem, "An outline of the dynamics of animal populations," *Australian Journal of Zoology* 2 (1954): 9–65; Kingsland, *Modeling Nature*, pp. 141–142.

90. Ian M. Mackerras, "Alexander John Nicholson," *Records of the Australian Academy of Science* 2, no. 1 (1970): 66–81; Carl Boris Schedvin, "Environment, economy, and Australian biology, 1890–1939," in *Scientific Colonialism: A Cross-cultural Comparison*, ed. Nathan Reingold and Marc Rothenberg (Washington: Smithsonian Institution Press, 1987), pp. 101–126, on pp. 120–123; idem, *Shaping Science and Industry: A History of Australia's Council for Scientific and Industrial Research, 1926–9* (Sydney: Allen & Unwin, 1987), pp. 46–56, 150–159; Carl B. Huffaker and Charles E. Kennett, "Biological control

of *Parlatoria oleae* (Colvée) through the compensatory action of two introduced parasites," *Hilgardia* 37 (1966): 283–335, on p. 330.

91. Paul DeBach and Harry S. Smith, "Are population oscillations inherent in the host-parasite relation?" *Ecology* 22 (1941): 363–369; idem, "The effect of host density on the rate of reproduction of entomophagous parasites," *Journal of Economic Entomology* 34 (1941): 741–745; idem, "Effects of parasite population density on rate of change of host and parasite populations," *Ecology* 28 (1947): 290–298; DeBach to Nicholson, 23 October 1947, PHD; Jessie Buchanan to Smith, 28 March 1939, HSS; Herbert G. Andrewartha and L. Charles Birch, *The Distribution and Abundance of Animals* (Chicago: University of Chicago Press, 1954), pp. 441–443, David L. Lack, *The Natural Regulation of Animal Numbers* (London: Oxford University Press, 1954), pp. 123–124.

92. For example, Huffaker, interview.

93. Smith, "Ecological aspects," pp. 25–26.

94. Smith, "Role of biotic factors," pp. 891–894.

95. Andrewartha and Birch, *Distribution and Abundance of Animals,* pp. 17–19; Huffaker, "Concept of balance in nature," p. 634; Smith, "Ecological aspects," pp. 12–13.

96. Smith, "Role of biotic factors," p. 890.

97. Carl B. Huffaker and Powers S. Messenger, "The concept and significance of natural control," in DeBach, ed., *Biological Control of Insect Pests and Weeds,* pp. 74–117, on p. 92; M. E. Solomon, untitled rejoinder to a letter by Alec Milne, *Nature* 182 (1958): 1252; Smith, "Role of biotic factors," pp. 885–890.

98. Smith, "Insect populations in relation to biological control," pp. 316–317; George C. Varley, "Ecology as an experimental science," *Journal of Animal Ecology* 26 (1957): 251–161; M. E. Solomon, "Meaning of density dependence and related terms in population dynamics," *Nature* 181 (1958): 1778–1780; Smith to Edward A. Steinhaus, 23 April 1954, EAS. On the debates over concepts and definitions, see also M. E. Solomon, "The natural control of animal populations," *Journal of Animal Ecology* 18 (1949): 1–35; Alec Milne, "Theories of natural control of insect populations," *Cold Spring Harbor Symposia on Quantitative Biology* 22 (1957): 253–271; idem, "Perfect and imperfect density dependence in population dynamics," *Nature* 182 (1958): 1251–1252; Solomon, rejoinder to Milne.

Chapter 5. SYSTEMATICS, EVOLUTION, AND APPLIED ENTOMOLOGY

1. John H. Perkins, *Insects, Experts, and the Insecticide Crisis: The Quest for New Pest Management Strategies* (New York: Plenum, 1981), pp. 33–36.

2. Stanley E. Flanders, "A new departure in codling moth control," *Journal of Economic Entomology* 18 (1925): 838–839; idem, "The mass production of *Trichogramma minutum* Riley and observations on the natural and artificial parasitism of the codling moth egg," *Transactions of the Fourth International Congress of Entomology* 2 (1929): 110–130; Smith to Leland O. Howard, 19 April 1927, HSS. For a discussion of the British arsenic scare, see James C. Whorton, *Before Silent Spring: Pesticides and Public Health in Pre-DDT America* (Princeton: Princeton University Press, 1974), pp. 133–175.

3. Stanley E. Flanders, "Mass production of egg parasites of the genus *Trichogramma*," *Hilgardia* 4 (1930): 465–501.

4. Smith to Howard, 23 October 1928, box 256, E35, RG7; Stanley E. Flanders, "Recent developments in *Trichogramma* production," *Journal of Economic Entomology* 23 (1930): 837–841; Harry S. Smith and Stanley E. Flanders, "Is *Trichogramma* becoming a fad?" ibid., 24 (1931): 666–672; Austin Winfield Morrill, "A discussion of Smith and Flanders' *Trichogramma* inquiry," ibid., 24 (1931): 1264–1263; Warren E. Hinds, B. A. Osterberger, and A. L. Dugas, "Sugar cane borer control by *Trichogramma* colonization in

Louisiana in 1932," ibid., 26 (1933): 758–767; Harold A. Jaynes and E. K. Bynum, "Experiments with *Trichogramma minutum* Riley as a control of the sugarcane borer in Louisiana," *U.S. Department of Agriculture Technical Bulletin* 743 (1941).

5. Stanley E. Flanders, "The temperature relationships of *Trichogramma minutum* as a basis for racial segregation," *Hilgardia* 5 (1931): 395–406; idem, "Habitat selection by *Trichogramma*," *Annals of the Entomological Society of America* 30 (1937): 208–210; William H. Thorpe, "Biological races in insects and allied groups," *Biological Reviews* 5 (1930): 177–212.

6. Stanley E. Flanders, "Identity of the common species of American *Trichogramma*," *Journal of Economic Entomology* 31 (1938): 456–457; Alvah Peterson, "How many species of *Trichogramma* occur in North America?" *Journal of the New York Entomological Society* 38 (1930): 1–8; Arthur B. Gahan to Flanders, 28 February 1931, box 94, E35, RG7; Gahan to Flanders, 1 February 1938, SEF.

7. Ernst Mayr, *The Growth of Biological Thought : Diversity, Evolution, and Inheritance* (Cambridge, Mass.: Harvard University Press, Belknap Press, 1982), pp. 266–276; Gordon F. Ferris, "The principles of systematic entomology," *Stanford University Publications, University Series, Biological Sciences* 5 (1928): 101–270, on pp. 134–158.

8. Pamela M. Henson, "Evolution and taxonomy: J. H. Comstock's research school in evolutionary entomology at Cornell University, 1874–1930" (Ph.D. dissertation, University of Maryland, College Park, 1990).

9. Alvah Peterson to Flanders, 26 September 1929, SEF.

10. Ernst Mayr, *Systematics and the Origin of Species* (1942; reprint ed., New York: Dover, 1964), pp. 151, 201–208; Alfred E. Emerson, "Termitophile distribution and quantitative characters as indicators of physiological speciation in British Guiana termites," *Annals of the Entomological Society of America* 28 (1935): 369–395; Hughes Evans, "European malaria policy in the 1920s and 1930s: The epidemiology of minutiae," *Isis* 80 (1989): 40–59, on pp. 54–56. Emerson identified sibling species of termites but joined other entomologists in considering biological races to be more common.

11. George Salt, "Experimental studies in insect parasitism. II. Superparasitism," *Proceedings of the Royal Society of London, Series B* 114 (1934): 455–476; idem, "The sense used by *Trichogramma* to distinguish between parasitized and unparasitized hosts," ibid., 122 (1937): 57–75; W. Robin Thompson, "Research on biological control," *Nature* 139 (1937): 552.

12. Vernon M. Stern, Evert I. Schlinger, and William R. Bowen, "Dispersal studies of *Trichogramma semifumatum* (Hymenoptera: Trichogrammatidae) tagged with radioactive phosphorus," *Annals of the Entomological Society of America* 58 (1965): 234–240; interview with Earl R. Oatman, 24 July 1987.

13. Stanley E. Flanders, "A biological phenomenon affecting the establishment of Aphelinidae as parasites," *Annals of the Entomological Society of America* 29 (1936): 251–255; idem, "Ovipositional instincts and developmental sex differences in the genus *Coccophagus*," *University of California Publications in Entomology* 6 (1937): 401–422, Thomas H. C. Taylor, "The campaign against *Aspidiotus destructor*, Sign., in Fiji," *Bulletin of Entomological Research* 20 (1935): 1–102, pp. 47–48; Stanley E. Flanders, "The practical application of biological studies of parasites employed in biological control," *Proceedings of the Sixth Pacific Science Congress* 4 (1939): 373–381.

14. Harry S. Smith, "Insect friends of the California farmer," typescript, dated 7 September 1932, HSS.

15. J. M. Thompson, "The orange industry—an economic study," *California Agricultural Experiment Station Bulletin* 622 (1938), p. 4; "Follow up the advantage" (editorial), *California Citrograph* 16 (1931): 449; Harry S. Smith and Harold Compere,

"Experiment station new insectary and biological control program," ibid., 16 (1931): 236–237.

16. B. A. O'Connor, "Biological control of insects and plants in Fiji," *Proceedings of the Seventh Pacific Science Congress* 4 (1953): 278–293; Taylor, "Campaign against *Aspidiotus destructor*"; idem, *The Biological Control of an Insect in Fiji: An Account of the Coconut Leaf-mining Beetle and its Parasite Complex* (London: Imperial Institute of Entomology, 1937).

17. Cyril E. Pemberton, "History of the Entomology Department Experiment Station, HSPA 1904–45," *Hawaiian Planters Record* 52, no. 1 (1948): 53–90; Philip H. Timberlake, "Biological control of insect pests in the Hawaiian Islands," *Proceedings of the Hawaiian Entomological Society* 6 (1927): 529–556, on p. 556; Augustus D. Imms, *Recent Advances in Entomology* (Philadelphia: P. Blakiston's Son & Co., 1931), pp. 339–342; Curtis P. Clausen, "Insect parasitism and biological control," *Annals of the Entomological Society of America* 29 (1936): 201–223, on p. 213.

18. Paul DeBach, "Successes, trends, and future possibilities," in *Biological Control of Insect Pests and Weeds*, ed. Paul DeBach (New York: Reinhold, 1964), pp. 675–712, on pp. 690–694.

19. Smith to H. Compere, 21 April 1932, HC.

20. Robert Boyce, "Insects and international relations: Canada, France, and British agricultural sanitary import restrictions between the Wars," *International History Review* 9 (1987): 1–27; Charles W. Woodworth, "Plant quarantine," typescript, *c.* 1929, EOE; Gordon F. Ferris, "The effectiveness of a plant quarantine," *Science* 71 (1930): 68–69; idem, "The plant quarantines once more," ibid., pp. 606–607; Harry S. Smith, "Plant quarantines, their aims, and their biological and economic justification," *California Department of Agriculture Bulletin* 19 (1930): 203–207; Harry S. Smith et al., "The efficiency and economic effects of plant quarantines in California," *California Agricultural Experiment Station Bulletin* 553 (1933), pp. 106–113; Smith to H. Compere, 19 July 1932, HC.

21. Harry R. Wellman, "Some economic aspects of regulating shipments of California oranges," *California Agricultural Experiment Station Circular* 338 (1936); Thompson, "Orange industry," pp. 39–40; Joseph G. Knapp, *The Advance of American Cooperative Enterprise* (Danville, Ill.: Interstate, 1973), pp. 229, 419–420; Clarke A. Chambers, *California Farm Organizations: A Historical Study of the Grange, the Farm Bureau, and the Associated Farmers, 1929–1941* (Berkeley: University of California Press, 1952), pp. 134–141; Charles C. Teague, *Ten Talks on Citrus Marketing* (Los Angeles, 1939).

22. Henry J. Quayle, "The development of resistance to hydrocyanic acid in certain scale insects," *Hilgardia* 11 (1938): 183–210.

23. Harold Compere, "Hazards of parasite hunting in foreign lands," *California Citrograph* 15 (1930): 533, 562–568; H. Compere to Smith, 18 March and 18 May 1930, HC.

24. Smith and Compere, "Experiment station new insectary"; "Search for scale parasites," *Citrus News* 7, no. 12 (1931): 13, 18; interview with Alfred M. Boyce, 13–14 May 1987.

25. Stanley E. Flanders, "Observations and experiences in searching for parasites in Australia," *Citrus Leaves* 12, no. 4 (1932): 15–18; idem, "The organization of biological control and its historical development," *Mededelingen van de Landbouwhogeschool en de Opzoekingsstations van den Staat te Gent* 20 (1955): 257–270, on p. 270; Smith to Flanders (telegram), 26 November 1931, SEF; Arthur W. Sampson and Kenneth W. Parker, "St. Johnswort on range lands of California," *California Agricultural Experiment Station Bulletin* 503 (1930): 19–20.

26. Leland O. Howard, "An interesting new genus and species of Encyrtidae," *Entomological News* 17 (1906): 121–122; Harold Compere and Harry S. Smith, "Notes on the life-history of two oriental chalcidoid parasites of *Chrysomphalus*," *University of California Publications in Entomology* 4 (1927): 63–73.

27. Harold Compere, "An appraisal of Silvestri's work in the Orient for the University of California, some misidentifications corrected, and two forms of *Casca* described as new species," *Bollettino del laboratorio di zoologia generale e agraria della facolta agraria in Portici* 33 (1953): 35–46, on p. 40; Harry S. Smith and Harold Compere, "An imported parasite attacks the yellow scale," *California Citrograph* 16 (1931): 328.

28. Harry S. Smith, "Biological control work," *California Department of Agriculture Bulletin* 12 (1923): 324–342; Harold Compere, "The red scale and its insect enemies," *Hilgardia* 31 (1961): 173–278, on pp. 186–189.

29. Compere, "Red scale and its insect enemies," pp. 224–232. Quotation is from H. Compere to Smith, 28 November 1932, HC.

30. Harry S. Smith, "University of California to continue red scale parasite search," *California Citrograph* 19 (1934): 263, 280–282; University of California, *Free Enterprise and University Research: A Record of Public Service by Business, Industry, Private Associations, and Foundations, through Support of Research Activities of the University of California* (Berkeley, 1954), p. 19; "Bread on the waters" (editorial), *California Citrograph* 19 (1934): 261.

31. Smith to Lee A. Strong, 18 May 1934, HC; Strong to Smith, 31 May 1934, HC; Smith to H. Compere, 5 February and 4 March 1935, HC.

32. Compere, "Red scale and its insect enemies," pp. 234–235; Harold Compere, "The insect enemies of the black scale, *Saissetia oleae* (Bern.), in South America," *University of California Publications in Entomology* 7 (1939): 75–90; Smith to H. Compere, 26 October 1934, HC.

33. "Biological control of red and black scale discussed by Professor Smith," *California Citrograph* 23 (1937): 59, 87; Compere, "Red scale and its insect enemies," pp. 238–239; H. Compere to Flanders, 8 January 1937, HC; Harry S. Smith, "Status of biological control of scale pests," *California Citrograph* 26 (1941): 58, 76–77.

34. Howard L. McKenzie, "Morphological differences distinguishing California red scale, yellow scale, and related species (Homoptera—Diaspididae)," *University of California Publications in Entomology* 6 (1937): 323–336; Compere, "Red scale and its insect enemies," p. 225; idem, "Appraisal of Silvestri's work," pp. 38–39.

35. Harold Compere, Stanley E. Flanders, and Harry S. Smith, "Use air transport from China for the introduction into California of a red scale inhabiting *Comperiella*," *California Citrograph* 26 (1941): 291, 300–301; Harry S. Smith, "A race of *Comperiella bifasciata* successfully parasitizes California red scale," *Journal of Economic Entomology* 35 (1942): 809–812; Compere, "Red scale and its insect enemies," pp. 264–269.

36. Howard, "Interesting new genus and species." Shortly afterward, Howard described the male in a separate paper, also on the basis of a single specimen sent from California. Leland O. Howard, "The male of *Comperiella*," *Entomological News* 18 (1907): 237.

37. Mayr, *Growth of Biological Thought*, p. 276; Julian Huxley, "Introductory: Towards the new systematics," in *The New Systematics*, ed. Julian Huxley (Oxford: Clarendon, 1940), pp. 1–46; William H. Thorpe, "Ecology and the future of systematics," ibid., pp. 341–364; John Smart, "Entomological systematics examined as a practical problem," ibid., pp. 475–492; Gordon F. Ferris, "The needs of systematic entomology," *Journal of Economic Entomology* 35 (1942): 732–738; Edward O. Essig, "The significance of taxonomy in the general field of economic entomology," ibid., 35 (1942): 739–743; Curtis P. Clausen, "The relation of taxonomy to biological control," ibid., 35 (1942): 744–748; C.

F. W. Muesebeck, "Fundamental taxonomic problems in quarantine and nursery inspection," ibid., 35 (1942): 753–758.

38. Theodore H. Frison, "The significance of economic entomology in the field of insect taxonomy," *Journal of Economic Entomology* 35 (1942): 749–752; Julian Huxley, *Evolution: The Modern Synthesis* (London: G. Allen & Unwin, 1942), pp. 299, 311–312, 471–473; Harry S. Smith, "Julian Huxley on evolution," *Ecology* 25 (1944): 477–479.

39. Harold Compere, "Collecting red and black scale parasites in Africa," *California Citrograph* 23 (1937): 58, 88–89.

40. Harold Compere, "Parasites of the black scale, *Saissetia oleae*, in Africa," *Hilgardia* 13 (1940): 387–425, pp. 388–394; Stanley E. Flanders, "*Metaphycus helvolus*, an encyrtid parasite of the black scale," *Journal of Economic Entomology* 35 (1942): 690–698; idem, "Competition and cooperation among parasitic Hymenoptera related to biological control," *Canadian Entomologist* 97 (1965): 409–422.

41. Smith to G. D. Hale Carpenter, 10 March 1939, HSS; Smith to H. Compere, 6 and 17 June 1939, HSS; Stanley E. Flanders, "Fig scale parasites introduced into California," *Journal of Economic Entomology* 50 (1957): 171–172; Smith to H. Compere, 20 April 1939, HC.

42. Much of the following discussion relies on interviews with former Smith students Laurence Jones (21 July 1987); Donald W. Clancy (22 July 1987); Janet Mabry Boyce (28 July 1987); Paul DeBach (9 July 1987); and Richard L. Doutt (30 May 1987).

43. Margaret W. Rossiter, *Women Scientists in America: Struggles and Strategies to 1940* (Baltimore: Johns Hopkins, 1982), pp. 225–241.

44. Paul DeBach and Harry S. Smith, "The effect of host density on the rate of reproduction of entomophagous parasites," *Journal of Economic Entomology* 34 (1941): 741–745; Harry S. Smith and Paul DeBach, "The measurement of the effect of entomophagous insects on population densities of their hosts," ibid., 35 (1942): 845–849; DeBach to Smith, 16 July 1945, PHD.

45. Thorpe, "Ecology and the future of systematics," pp. 342–344.

46. Ernst Mayr, "The role of systematics in the evolutionary synthesis," in *The Evolutionary Synthesis: Perspectives on the Unification of Biology*, ed. Ernst Mayr and William B. Provine (Cambridge, Mass.: Harvard University Press, 1980), pp. 123–136, on p. 128; William Bateson, *Problems of Genetics* (New Haven: Yale University Press, 1913), pp. 236–242; Theodosius Dobzhansky, "*Drosophila miranda*, a new species," *Genetics* 20 (1935): 377–391; idem, "A critique of the species concept in biology," *Philosophy of Science* 2 (1935): 344–355; idem, *Genetics and the Origin of Species* (New York: Columbia University Press, 1937).

47. Harry S. Smith, "Racial segregation in insect populations and its significance in applied entomology," *Journal of Economic Entomology* 34 (1941): 1–13, on pp. 1–2; Robert C. Dickson, "Inheritance of resistance to hydrocyanic acid fumigation in the California red scale," *Hilgardia* 13 (1941): 513–521.

48. Dobzhansky, *Genetics and the Origin of Species*, p. 161.

49. Smith, "Racial segregation," p. 5; Dobzhansky, *Genetics and the Origin of Species*, pp. 228–258.

50. Philip H. Timberlake to H. Morton Armitage, 30 June 1930, PHT; A. M. Boyce, interview.

51. Benjamin D. Walsh, "On phytophagic varieties and phytophagic species," *Proceedings of the Entomological Society of Philadelphia* 3 (1864): 403–430.

52. Frank C. Craighead, "Hopkins host-selection principle as related to certain cerambycid beetles," *Journal of Agricultural Research* 22 (1921): 189–220; idem, "The host-selection principle as advanced by Walsh," *Canadian Entomologist* 55 (1923): 76–80.

53. Thorpe, "Biological races in insects and allied groups."

54. Smith, "Race of *Comperiella* parasitizes red scale," p. 811; idem, "Racial segregation," p. 2.

55. Harry S. Smith, "On the life-history and successful introduction into the United States of the Sicilian mealy-bug parasite," *Journal of Economic Entomology* 10 (1917): 262–268; idem, review of *The Biological Control of Insects* by Harvey L. Sweetman, ibid., 30 (1937): 218–220; Smith, "Racial segregation," p. 8; Richard W. Burkhardt, Jr., "Lamarckism in Britain and the United States," in Mayr and Provine, eds., *The Evolutionary Synthesis*, pp. 343–352; Peter J. Bowler, *The Eclipse of Darwinism: Anti-Darwinian Evolution Theories in the Decades around 1900* (Baltimore: Johns Hopkins University Press, 1983), p. 60; William H. Thorpe and F. G. W. Jones, "Olfactory conditioning in a parasitic insect and its relation to the problem of host selection," *Proceedings of the Royal Society of London, Series B* 126 (1937): 56–81; Thorpe, "Ecology and the future of systematics," p. 354.

56. Mayr, "Role of systematics," pp. 129–133; Dobzhansky, *Genetics and the Origin of Species*, pp. 256–257; Smith, "Racial segregation," 7–8; Thorpe, "Ecology and the future of systematics," p. 355; Mayr, *Systematics and the Origin of Species*, pp. 192–215.

57. Clausen, "Insect parasitism and biological control," pp. 202–203; Smith, "Racial segregation," pp. 3–7.

58. A. Wilkes, "The effects of selective breeding on the laboratory propagation of insect parasites," *Proceedings of the Royal Society of London, Series B* 134 (1947): 227–245; Paul DeBach, "Selective breeding to improve adaptations of parasitic insects," *Proceedings of the Tenth International Congress of Entomology* 4 (1958): 759–768.

59. Jan Sapp, *Beyond the Gene: Cytoplasmic Inheritance and the Struggle for Authority in Genetics* (New York: Oxford University Press, 1987), pp. 45–46.

60. Smith, "Racial segregation," p. 4 (italics original).

61. Theodosius Dobzhansky, *Genetics and the Origin of Species*, 2nd ed. (New York: Columbia University Press, 1941), pp. 190–193.

62. Smith, "Racial segregation," pp. 9–12 (quotation from p. 9); idem, "Insect populations in relation to biological control," *Ecological Monographs* 9 (1939): 311–320.

63. Smith, "Racial segregation," p. 9.

64. Harry S. Smith, "Present status of biological control of mealybugs," *California Citrograph* 14 (1929): 428–429; idem, "Economic entomology and the national welfare," *Journal of Economic Entomology* 33 (1940): 587–588; idem, "Status of biological control" (1941), p. 58.

65. Harry S. Smith, "The biological control program in relation to California agriculture," *California Citrograph* 31 (1946): 414–415, 452–453.

Chapter 6. GROWTH AND CONFLICT IN THE AGE OF DDT

1. For a history of the controversy over DDT, see Thomas R. Dunlap, *DDT: Scientists, Citizens, and Public Policy* (Princeton: Princeton University Press, 1981).

2. Jim Hightower, *Hard Tomatoes, Hard Times* (Cambridge, Mass.: Shenkman, 1973); John H. Perkins, "Insects, food, and hunger: The paradox of plenty for U.S. entomology," *Environmental Review* 7 (1983): 71–86.

3. Harold Compere, "A systematic study of the genus *Aphytis* Howard (Hymenoptera, Aphelinidae) with descriptions of new species," *University of California Publications in Entomology* 10 (1955): 271–320, on pp. 271–272; idem, "The role of systematics in biological control: A backward look," *Israel Journal of Entomology* 4 (1969): 5–10.

4. "The picture changes" and "No excuse for this" (editorials), *California Citrograph*

27 (1942): 65; Harold Compere, "Your health—my oranges," *Citrograph* 55 (1971): 324-327; idem, "The search for parasites of red scale and other citrus pests in Brazil," *Citrus Leaves* 15 (1935): 8-9; H. Compere to Smith, 22 May 1947, HC; H. Compere to R. G. Nel, 5 January 1946, HC.

5. David B. Mackie, "One year with the oriental fruit moth," *California Department of Agriculture Bulletin* 33 (1944): 4-17; Curtis P. Clausen, "Olethreutidae," in *Introduced Parasites and Predators of Arthropod Pests and Weeds: A World Review,* ed. Curtis P. Clausen, Agriculture Handbook no. 480 (Washington: U.S. Department of Agriculture, 1978), pp. 210-222, on pp. 213-216; Harry S. Smith, "Cost of producing *Macrocentrus* by the potato tuber-worm method," *Journal of Economic Entomology* 38 (1945): 316-319; Glenn L. Finney, Stanley E. Flanders, and Harry S. Smith, "Mass culture of *Macrocentrus ancylivorus* and its host, the potato tuber moth," *Hilgardia* 17 (1947): 437-483.

6. Robert C. Dickson, "Factors governing the induction of diapause in the oriental fruit moth," *Annals of the Entomological Society of America* 42 (1949): 511-537.

7. Hutchison to Robert G. Sproul, 28 December 1938, 1938: 360, UCA-PF.

8. Paolo S. A. Palladino, "Entomology and ecology: The ecology of entomology" (Ph.D. dissertation, University of Minnesota, 1989), pp. 10, 69-73.

9. Dunlap, *DDT,* pp. 35-36; William B. Herms, "An analysis of some of California's major entomological problems," *Journal of Economic Entomology* 19 (1926): 262-270; Ray F. Smith, "The origins of integrated control in California, an account of the contributions of Charles W. Woodworth," *Pan-Pacific Entomologist* 50 (1974): 426-440; Edward O. Essig and Abraham E. Michelbacher, "The alfalfa weevil," *California Agricultural Experiment Station Bulletin* 567 (1933); William M. Hoskins, A. D. Borden, and Abraham E. Michelbacher, "Recommendations for a more discriminating use of insecticides," *Proceedings of the Sixth Pacific Science Congress* 6 (1939): 119-123; Abraham E. Michelbacher, "Some entomological observations in California," *Journal of Economic Entomology* 33 (1940): 141-143.

10. Smith to Richard L. Doutt, 12 August 1949, PHD; interview with Richard A. Smith, 30 May 1987; interview with Alfred M. Boyce, 13-14 May 1987.

11. Smith to W. D. Drew, 10 June 1939, HSS; interview with Paul DeBach, 9 July 1987; interview with Richard L. Doutt, 30 May 1987; interview with Theodore W. Fisher, 28 May 1987; interview with Carl B. Huffaker, 10 June 1987; Harry S. Smith, "Production of parasites for oriental fruit moth control," typescript, ca. 1944, Harry S. Smith et al. file, UCR-BC.

12. Edward A. Steinhaus, *Disease in a Minor Chord* (Columbus: Ohio State University Press, 1975), pp. 60-147; Harry S. Smith, "The use of fungous diseases against the black scale," *California State Commission of Horticulture Bulletin* 4 (1915): 109-111; Richard C. Sawyer, "Monopolizing the insect trade: Biological control in the USDA, 1888-1951," *Agricultural History* 64, no. 2 (1990): 271-285, on p. 279.

13. Steinhaus, *Disease in a Minor Chord,* pp. 147-171.

14. Smith to Alfred B. Baird, 4 February 1946, EAS.

15. This is a recurring theme in Steinhaus's largely autobiographical book, *Disease in a Minor Chord.*

16. Edward A. Steinhaus, "Possible use of *Bacillus thuringiensis* Berliner as an aid in the biological control of the alfalfa caterpillar," *Hilgardia* 20 (1951): 359-381; idem, *Disease in a Minor Chord,* pp. 167-214; idem, "Potentialities for microbial control of insects," *Journal of Agricultural and Food Chemistry* 4 (1956): 676-680; Irvin M. Hall, "Use of micro-organisms in biological control," in *Biological Control of Insect Pests and Weeds,* ed. Paul DeBach (New York: Reinhold, 1964), pp. 610-628.

17. R. C. L. Perkins and Otto H. Swezey, "The introduction into Hawaii of insects that

attack lantana," *Hawaiian Sugar Planters' Association, Entomological Series, Bulletin* 16 (1924); C. J. Davis and Noel L. H. Krauss, "Recent developments in the biological control of weed pests in Hawaii," *Proceedings of the Hawaiian Entomological Society* 18 (1962): 65–67; Alan P. Dodd, *The Biological Campaign against Prickly Pear* (Brisbane: Commonwealth Prickly Pear Board, 1940); Frank Wilson, "The entomological control of St. John's Wort (*Hypericum perforatum* L.), with particular reference to the insect enemies of the weed in southern France," *Australian Council of Scientific and Industrial Research Bulletin* 169 (1943).

18. Sawyer, "Monopolizing the insect trade," p. 284; Arthur W. Sampson and Kenneth W. Parker, "St. Johnswort on range lands of California," *California Agricultural Experiment Station Bulletin* 503 (1930); Arthur W. Sampson and Harry E. Malmsten, "Stock-poisoning plants in California," ibid., 593 (1935); Smith to Charles L. Marlatt, 1 February 1932, SE65, RG7; Lee A. Strong to Smith, 18 February 1932, ibid.; Marlatt to Smith, 3 March 1932, ibid.; Huffaker, interview.

19. Richard D. Goeden, Charles A. Fleschner, and Donald W. Ricker, "Biological control of prickly pear cacti on Santa Cruz Island, California," *Hilgardia* 38 (1967): 579–606; Smith to David B. Mackie, 10 June 1939, HSS; Smith to Percy N. Annand, 9 September 1942, box 1465, SE22, RG7. The mealybug was introduced from Hawaii, which got it from Australia, which got it from Mexico. Its native range was later found to extend within 40 miles of Riverside.

20. Clausen to Smith, 24 April 1944, box 167, SE8, RG7.

21. Huffaker, interview; Doutt, interview. Huffaker recalled that he knew he would get the job, because no one else had the combination of qualifications Smith wanted.

22. Clausen to Holloway, 17 April 1943, box 1061, SE121, RG7; Clausen to Holloway, 25 October 1944, box 167, SE8, RG7.

23. James K. Holloway and Carl B. Huffaker, "Insects to control a weed," *U.S. Department of Agriculture Yearbook* 1952: 135–140; Carl B. Huffaker, "Quantitative studies on the biological control of St. John's wort (Klamath weed) in California," *Proceedings of the Seventh Pacific Science Congress* 4 (1953): 303–313; Carl B. Huffaker and Charles E. Kennett, "A ten-year study of vegetational changes associated with biological control of Klamath weed," *Journal of Range Management* 12 (1959): 69–82.

24. Robert P. McIntosh, *The Background of Ecology: Concept and Theory* (Cambridge: Cambridge University Press, 1985), pp. 87–88, 117–118; Frederic E. Clements and Victor E. Shelford, *Bio-Ecology* (New York: John Wiley & Sons, 1939), pp. 285–293; Arthur W. Sampson, "Plant succession in relation to range management," *U.S. Department of Agriculture Bulletin* 791 (1919). Regarding Sampson's use of succession theory, see C. H. Wasser, Early development of technical range management, ca. 1895–1945," *Agricultural History* 51 (1977): 63–77; and Maarten Heyboer, "Grass-counters, stock-feeders, and the dual orientation of range science, 1895–1960" (Ph.D. dissertation, Virginia Polytechnic Institute and State University, 1992), pp. 135–209.

25. Wilson, "Entomological control of St. John's Wort," pp. 63–66; idem, "Some aspects of the control of weeds by insects," *Proceedings of the Seventh Pacific Science Congress* 3 (1953): 294–299.

26. On the persistence of Clements's ideas, see Joel B. Hagen, *An Entangled Bank: The Origins of Ecosystem Ecology* (New Brunswick: Rutgers University Press, 1992), especially pp. 31–49.

27. Carl B. Huffaker, "The return of native perennial bunch-grass following the removal of Klamath weed (*Hypericum perforatum* L.) by imported beetles," *Ecology* 32 (1951): 443–458; idem, "Quantitative studies on St. John's wort"; idem, "Fundamentals of biological control of weeds," *Hilgardia* 27 (1957): 101–157, on pp. 123–137; idem, "Some

concepts on the ecological basis of biological control of weeds," *Canadian Entomologist* 94 (1962): 507–514; idem, interview.

28. Carl. B. Huffaker, "The concept of balance in nature," *Proceedings of the Tenth International Congress of Entomology* 2 (1958): 625–636; idem, "Biological control of weeds with insects," *Annual Review of Entomology* 4 (1959): 251–276, on pp. 251–252; Carl B. Huffaker and Powers S. Messenger, "Population ecology—historical development," in DeBach, ed., *Biological Control of Insect Pests and Weeds.* pp. 45–73; idem, "The concept and significance of natural control," ibid., pp. 74–117.

29. Sharon E. Kingsland, *Modeling Nature: Episodes in the History of Population Ecology* (Chicago: University of Chicago Press, 1985), pp. 171–174; Carl B. Huffaker, "Experimental studies on predation: Dispersion factors and predator-prey oscillations," *Hilgardia* 27 (1958): 343–383; Carl B. Huffaker and Charles E. Kennett, "Experimental studies on predation: Predation and cyclamen-mite populations on strawberries in California," ibid., 26 (1956): 191–222; Herbert G. Andrewartha and L. Charles Birch, "Some recent contributions to the study of the distribution and abundance of insects," *Annual Review of Entomology* 5 (1960): 219–242, on pp. 232–234.

30. Alfred M. Boyce, *Odyssey of an Entomologist: Adventures on the Farm, at Sea, and in the University* (Riverside, Calif.: UC Riverside Foundation, 1987), pp. 141–143, Alfred M. Boyce and William H. Ewart, "DDT for control of citrus thrips and citricola scale in central California," *California Citrograph* 31 (1946): 240–241; Russell S. Woglum, "Insect pest control report, 1945–6," Field Services: Pest Control Department, SGA; Paul DeBach, "Cottony cushion scale, vedalia and DDT in central California," *California Citrograph* 32 (1947): 406–407; Paul DeBach to W. B. Saunders, 15 January 1948, PHD.

31. Boyce, interview; "Concerning the accidental death of Keith E. Hughes from exposure to parathion," mimeograph, Citrus Experiment Station Division of Entomology, 12 September 1949, PHD; Boyce, *Odyssey of an Entomologist,* pp. 246–247; James T. Griffiths, Charles R. Stearns, Jr., and William L. Thompson, "Parathion hazards encountered spraying citrus in Florida," *Journal of Economic Entomology* 44 (1951): 160–163.

32. Paul DeBach, "An insecticidal check method for measuring the efficacy of entomophagous insects," *Journal of Economic Entomology* 39 (1946): 695–697.

33. H. Compere to Richard H. LePelley, 19 August 1946, HC; Harold Compere, "The red scale and its insect enemies," *Hilgardia* 31 (1961): 173–278, on pp. 241–251; Gerald C. Ullyett, "Insecticide spray programs and biological control in South Africa," *Journal of Economic Entomology* 41 (1948): 337–339; H. Compere to Smith, 22 May 1947, HC.

34. "Entomological explorer returns," *California Citrograph* 34 (1948): 70–71; H. Compere to DeBach, 20 October 1947, PHD; Smith to Compere, 8 July 1947, HC; H. Compere to Smith, 20 September and 24 October 1947, HC. Boyce recalled that this incident ended the close friendship between Compere and Smith and prompted Compere to take the university out of his will. Others failed to corroborate Boyce, however, and Smith's surviving children complained about Boyce's story. By most accounts, Compere and Smith remained nearly as close as ever and were still frequent guests in each other's homes. Although they exchanged harsh words initially, a friendly tone returned to their letters even before Compere returned to California in the spring of 1948. Boyce, *Odyssey of an Entomologist,* pp. 251–253; Boyce, Doutt, Richard Smith, interviews; interview with Lawrence F. Smith, 27 May 1987; interview with Laurence Jones, 21 July 1987.

35. Paul DeBach, Charles A. Fleschner, and Everett J. Dietrick, "Natural control of the red scale on citrus," *California Citrograph* 34 (1948): 6, 38–39; idem, "Natural control of the California red scale in untreated citrus orchards in southern California," *Proceedings of the Seventh Pacific Science Congress* 4 (1953): 236–249; Paul DeBach, "Biological

control of red scale in San Diego County," *California Citrograph* 37 (1952): 136–137, 158–160; Harry S. Smith and Stanley E. Flanders, "The search for natural enemies of citrus pests," ibid., 35 (1950): 362, 376–378.

36. Paul DeBach, Everett W. Dietrick, and Charles A. Fleschner, "A new technique for evaluating the efficiency of entomophagous insects in the field," *Journal of Economic Entomology* 42 (1949): 546; Paul DeBach, Charles A. Fleschner, and Everett W. Dietrick, "A biological check method for evaluating the effectiveness of entomophagous insects," ibid., 44 (1951): 763–766.

37. Clay Lyle, "Achievements and possibilities in pest eradication," *Journal of Economic Entomology* 40 (1947): 1–8.

38. James C. Leary, William I. Fishbein, and Lawrence C. Salter, *DDT and the Insect Problem* (New York: Macmillan, 1947), p. 9.

39. E. H. Strickland, "Could the widespread use of DDT be a disaster?" *Entomological News* 56 (1945): 85–88; Abraham E. Michelbacher, "The importance of ecology in insect control," *Journal of Economic Entomology* 38 (1945): 129–130; Harry S. Smith, "Racial segregation in insect populations and its significance in applied entomology," ibid., 34 (1941): 1–13.

40. DeBach form letters to various entomologists, 16 November 1948 and 27 July 1949, PHD; DeBach to Clausen, 1 November 1949, PHD; Nominations for committee, PHD; Paul DeBach, "The necessity for an ecological approach to pest control on citrus in California," *Journal of Economic Entomology* 44 (1951): 443–447; Gerald C. Ullyett, "Insects, man, and the environment," ibid., pp. 459–464; James T. Griffiths, "Possibilities for better citrus insect control through the study of the ecological effects of spray programs," ibid., pp. 464–468.

41. Compere to Stanley E. Flanders, 8 July 1931, HC.

42. Boyce, *Odyssey of an Entomologist,* pp. 47–215; Boyce, interview.

43. Paul DeBach, *Biological Control by Natural Enemies* (London: Cambridge University Press, 1974), pp. 169–171; Charles E. Kennett, Carl B. Huffaker, and Glenn L. Finney, "The role of an autoparasitic aphelinid, *Coccophagoides utilis* Doutt, in the control of *Parlatoria oleae* (Colvée)," *Hilgardia* 37 (1966): 255–282.

44. Paul DeBach, "Purple scale parasites in southern California," *California Citrograph* 38 (1953): 219–222.

45. Smith and Flanders, "Search for natural enemies"; Stanley E. Flanders, "Hymenopterous parasites of three species of oriental scale insects," *Bollettino del Laboratorio di zoologia generale e agraria della Facolta agraria in Portici* 33 (1953): 10–28; idem, "Some biological control aspects of taxonomy exemplified by the genus *Aphytis* (Hymenoptera: Aphelinidae)," *Canadian Entomologist* 96 (1964): 888–893; Compere, "Systematic study of genus *Aphytis.*"

46. Richard L. Doutt, "An evaluation of some natural enemies of the olive scale," *Journal of Economic Entomology* 47 (1954): 39–43; David Rosen and Paul DeBach, "Diaspididae," in Clausen, ed., *Introduced Parasites and Predators,* pp. 78–128, on pp. 114–117; Ernst Mayr, E. Gorton Linsley, and Robert L. Usinger, *Methods and Principles of Systematic Zoology* (New York: McGraw-Hill, 1953); Mostafa Hafez and Richard L. Doutt, "Biological evidence of sibling species in *Aphytis maculicornis* (Masi) (Hymenoptera, Aphelinidae)," *Canadian Entomologist* 86 (1954): 90–96; Boyce, *Odyssey of an Entomologist,* pp. 214–215.

47. Paul DeBach and Ragnhild A Sundby, "Competitive displacement between ecological homologues," *Hilgardia* 34 (1963): 105–166; Paul DeBach and John Landi, "The introduced purple scale parasite, *Aphytis lepidosaphes* Compere, and a method of integrating chemical with biological control," ibid., 31 (1961): 459–497. *A. lepidosaphes* proved even

more valuable elsewhere, often arriving by accident. Just as entomologists in Florida were about to have the parasite sent from California, they suddenly found it already on the way to complete control of purple scale. They never knew how the parasite got to Florida. Donald W. Clancy and Martin H. Muma, "Purple scale parasite found in Florida," *Journal of Economic Entomology* 52 (1959): 1025–1026; interview with Donald W. Clancy, 22 July 1987.

48. Paul DeBach, David Rosen, and Charles E. Kennett, "Biological control of coccids by introduced natural enemies," in *Biological Control,* ed. Carl B. Huffaker (New York: Plenum, 1971), pp. 165–194, on p. 177; DeBach, *Biological Control by Natural Enemies,* pp. 171–181; Richard L. Doutt, "A taxonomic analysis of parasitic Hymenoptera reared from *Parlatoria oleae* (Colvée)," *Hilgardia* 37 (1966): 219–231; S. W. Broodryk and Richard L. Doutt, "The biology of *Coccophagoides utilis* Doutt (Hymenoptera, Aphelinidae)," ibid., pp. 233–254; Carl B. Huffaker and Charles E. Kennett, "Biological control of *Parlatoria oleae* (Colvée) through the compensatory action of two introduced parasites," ibid., pp. 283–335; DeBach and Sundby, "Competitive displacement," pp. 162–163.

49. Stanley E. Flanders, "George Compere, pioneer in the biological control of red scale," *California Citrograph* 34 (1949): 160–162; Compere, "Red scale and its insect enemies," pp. 204–207; Paul DeBach, "The importance of taxonomy to biological control as illustrated by the cryptic history of *Aphytis holoxanthus* n. sp. (Hymenoptera: Aphelinidae), a parasite of *Chrysomphalus aonidum,* and *Aphytis coheni* n. sp., a parasite of *Aonidiella aurantii,*" *Annals of the Entomological Society of America* 53 (1960): 701–705.

50. Boyce, *Odyssey of an Entomologist,* pp. 215–219; idem, interview; DeBach, interview; Paul DeBach and David Rosen, *Biological Control by Natural Enemies,* 2nd ed. (Cambridge: Cambridge University Press, 1991), pp. 221–222.

51. Boyce, interview; Huffaker, interview; University of California, *Free Enterprise and University Research: A Record of Public Service by Business, Industry, Private Associations, and Foundations through Support of the Research Activities of the University of California* (Berkeley, 1954), pp. 20–23. Chemical grants are listed in the annual, later biennial, reports of the California Agricultural Experiment Station.

52. Harry S. Smith, "The biological control program in relation to California agriculture," *California Citrograph* 51 (1946): 414–415, 452–453; Steinhaus, *Disease in a Minor Chord,* pp. 370–371; Smith to Steinhaus, 13 and 21 December 1945, EAS.

53. Quotation from an unpublished obituary apparently written by Robert van den Bosch, in Smith file, UCR-BC.

54. Steinhaus, *Disease in a Minor Chord,* pp. 218–220; Boyce, interview; interview with Ray F. Smith, 13 June 1987. Virtually all of Harry Smith's former students and colleagues interviewed mentioned his intense interest in football.

55. Fisher, interview; Doutt, interview; Steinhaus to Smith, 4 August 1949, EAS.

56. Smith to Clausen, 1 March 1951, Curtis P. Clausen file, UCR-DE.

57. H. Morton Armitage, "The oriental fruit fly from the mainland viewpoint," *Journal of Economic Entomology* 42 (1949): 713–716; Curtis P. Clausen, Donald W. Clancy, and Q. C. Chock, "Biological control of the oriental fruit fly *(Dacus dorsalis* Hendel) and other fruit flies in Hawaii," *U.S. Department of Agriculture Technical Bulletin* 1322 (1965).

58. Clausen to Avery S. Hoyt, 11 December 1950, box 159, SE8, RG7; James I. Hambleton and Theodore R. Gardner to Hoyt, 1 November 1951, ibid.

59. Clausen to Hutchison, 4 December 1950, Clausen file, UCR-DE; Clausen to Robert G. Sproul, 9 April 1951, ibid.

60. Smith to Steinhaus, ? February 1955, EAS; Boyce, Clancy, Jones, Huffaker, Ray Smith, Richard Smith, interviews.

61. Smith to Clausen, 28 March 1951, Clausen file, UCR-DE; Hutchison to Clausen, 7 March 1951, ibid.; Gordon S. Watkins to Clausen, 29 November 1955, ibid.; Smith to Leon D. Batchelor, 30 October 1950, Stanley E. Flanders file, UCR-DE.

62. Verne A. Stadtman, ed., *The Centennial Record of the University of California* (Berkeley: University of California Printing Dept., 1967), pp. 85, 153, 167, 433, 438; Verne A. Stadtman, *The University of California, 1868–1968* (New York: McGraw-Hill, 1970), pp. 350–405.

63. Huffaker to Biological Control Coordinating Committee (Smith, Flanders, DeBach, Steinhaus, Doutt), 27 January 1950, PHD; Smith to Huffaker, 6 February 1950, EAS.

64. J. Linsley Gressitt and Stanley E. Flanders, "New developments in the transport of beneficial insects," *Journal of Economic Entomology* 42 (1949): 154; Blair R. Bartlett, "The international shipment of adult entomophagous insects," *Annals of the Entomological Society of America* 55 (1962): 448–455; DeBach, *Biological Control by Natural Enemies,* pp. 155–157.

65. Huffaker, Doutt, DeBach, Boyce, interviews; interview with Kenneth S. Hagen, 19 June 1987.

66. Flanders to H. Compere, 12 June and 23 October 1947, HC; Fisher, interview; DeBach, interview; H. Compere letter addressed to Alec Milne, 22 October 1957, marked "Not sent at request of DeBach & Fleschner," HC; H. Compere to F. M. Coray, 8 September 1950, HC; Richard Martinez, "University makes neighborhood go," *Riverside Press-Enterprise,* 22 October 1986.

67. DeBach to Flanders, 27 September 1950, PHD; Stanley E. Flanders, "Mass culture of California red scale and its golden chalcid parasites," *Hilgardia* 21 (1951): 1–42, on p. 40; idem, "*Casca*'s elusive husband," *California Citrograph* 39 (1954): 343, 352; Stanley E. Flanders, J. Linsley Gressitt, and Theodore W. Fisher, "*Casca chinensis,* an internal parasite of California red scale," *Hilgardia* 28 (1958): 65–91; Paul DeBach and Ernest B. White, "Commercial mass culture of the commercial red scale parasite, *Aphytis lingnanensis,*" *California Agricultural Experiment Station Bulletin* 770 (1960), pp. 5–6; DeBach, interview; interview with Charles E. Kennett, 17 June 1987; interview with Earl R. Oatman, 24 July 1987.

68. Steinhaus, *Disease in a Minor Chord,* pp. 233–234; Doutt, interview; Huffaker, interview; Steinhaus to Clausen, 20 March 1956, EAS; Doutt, Glenn L. Finney, Kenneth S. Hagen, Huffaker, Mauro E. Martignoni, Steinhaus, Yoshinori Tanada, and A. D. Telford to Harry R. Wellman, 23 September 1957, Clausen file, UCR-DE; Comments submitted by Flanders and H. Compere on the above letter, ibid. Compere marked the comments, "Written by H. S. Smith Sept. 31, 1957." An earlier draft of the group letter requesting a separate department, dated 19 September 1957, included Clausen's name. Steinhaus's copy, marked "This must have been prepared by Doutt for Clausen's signature," is in EAS.

69. Vernon M. Stern et al., "The integrated control concept," *Hilgardia* 29 (1959): 81–101; John H. Perkins, *Insects, Experts, and the Insecticide Crisis: The Quest for New Pest Management Strategies* (New York: Plenum, 1981).

70. Thomas H. C. Taylor, "Biological control of insect pests," *Annals of Applied Biology* 42 (1955): 190–196.

71. Michelbacher, "Ecology in insect control," p. 130; "Supervised pest control program of field crop growers," *California Citrograph* 34 (1949): 314–315; Ray F. Smith and William W. Allen, "Insect control and the balance of nature," *Scientific American* 190, no. 6 (1954): 38–42; Perkins, *Insects, Experts, and the Insecticide Crisis,* pp. 76–78; Hagen, interview.

72. Blair R. Bartlett, "Relative toxicity of field-weathered insecticide residues to entomophagous insects associated with citrus in California," *Journal of Economic*

Entomology 46 (1953): 565–569; idem, "Natural predators: Can selective insecticides help to preserve biotic control?" *Agricultural Chemicals* 11, no. 2 (1956): 42–44, 107–109; idem, "Where are we going in biological control?" *California Citrograph* 42 (1957): 339, 360–361. Ray F. Smith once claimed to have been the first to use the phrase "integrated control" in print, in 1954. Michelbacher used the phrase in a 1952 paper, only in passing and with no indication that a label was in the making. Whether Bartlett picked up "integrated control" there or coined it independently is unknown, but its standardization seems to come from his usage beginning in 1953. See R. F. Smith, "Origins of integrated control," p. 426; Abraham E. Michelbacher and Oscar G. Bacon, "Walnut insect and spider-mite control in northern California," *Journal of Economic Entomology* 45 (1952): 1020–1057, on p. 1020.

73. A. D. Pickett et al., "The influence of spray programs on the fauna of apple orchards in Nova Scotia. I. An appraisal of the problem and a method of approach," *Scientific Agriculture* 26 (1946): 590–600; A. D. Pickett, "The philosophy of orchard insect control," *Annual Report of the Entomological Society of Ontario* 79 (1948): 37–41; idem, "A critique on insect chemical control methods," *Canadian Entomologist* 81 (1949): 67–76; A. D. Pickett, William L. Putman, and E. J. LeRoux, "Progress in harmonizing biological and chemical control of orchard pests in eastern Canada," *Proceedings of the Tenth International Congress of Entomology* 3 (1958): 169–174.

74. John H. Perkins, "The quest for innovation in agricultural entomology, 1945–1978," in *Pest Control: Cultural and Environmental Aspects,* ed. David Pimentel and John H. Perkins (Boulder, Colo.: Westview Press, 1980), pp. 23–80.

75. A. D. Pickett and N. A. Patterson, "The influence of spray programs on the fauna of apple orchards in Nova Scotia. IV. A review," *Canadian Entomologist* 85 (1953): 472–478; Stern et al., "Integrated control concept," pp. 87–96; Ray F. Smith and Kenneth S. Hagen, "Integrated control programs in the future of biological control," *Journal of Economic Entomology* 52 (1959): 1106–1108.

76. Thomas R. Dunlap, "Science as a guide in regulating technology: The case of DDT in the United States," *Social Studies of Science* 8 (1978): 265–285; idem, *DDT.*

77. DeBach, *Biological Control by Natural Enemies,* pp. 2–20; Stern et al., "Integrated control concept," p. 94; Kennett, interview; Compere, "Your health—my oranges."

78. Robert van den Bosch, *The Pesticide Conspiracy* (Garden City, N.Y.: Doubleday, 1978); van den Bosch to DeBach, 17 May 1976, PHD; Richard L. Doutt and Ray F. Smith, "The pesticide syndrome—diagnosis and suggested prophylaxis," in Huffaker, ed., *Biological Control,* pp. 3–15, on pp. 5–6; Ray Smith, Kennett, DeBach, Doutt, Boyce, interviews.

79. Paul DeBach, "Successes, trends, and future possibilities," in DeBach, ed., *Biological Control of Insect Pests and Weeds,* pp. 673–713, on pp. 711–712; "Entomology notes," *Citrograph* 55 (1970): 184.

80. James T. Griffiths, "Where do we go from here?" *Florida Entomologist* 36 (1953): 135–140; idem, "Possibilities for better citrus insect control"; Oscar E. Anderson, Jr., *Refrigeration in America: A History of a New Technology and its Impact* (Princeton: Princeton University Press, 1953), pp. 201–203; Thomas A. Rector, "Frozen concentrated orange juice—its research background," *Refrigerating Engineering* 58 (1950): 349–353; James T. Griffiths, and William L. Thompson, "Reduced spray programs for citrus for canning plants in Florida," *Journal of Economic Entomology* 46 (1953): 930–936; Ronald W. Ward and Richard L. Kilmer, *The Citrus Industry: A Domestic and International Economic Perspective* (Ames: Iowa State University Press, 1989), p. 9.

81. Steinhaus, *Disease in a Minor Chord,* p. 235; Doutt, Jones, Boyce, interviews; interview with Charles A. Fleschner, 4 June 1987. Apparently there was already some fear in Riverside that if Fleschner did not accept the chair, the department would be split and merged with entomology on the respective campuses. Lorenne Sisson to Fleschner, 29 May

1959, Charles A. Fleschner file, UCR-DE.

82. Charles A. Fleschner, "Field approach to population studies of tetranychid mites on citrus and avocado in California," *Proceedings of the Tenth International Congress of Entomology* 2 (1958): 669–674; idem, "Natural enemies of tetranychid mites on citrus and avocado in southern California," ibid., 4 (1958): 627–631; idem, interview.

83. Boyce, Fleschner, Hagen, interviews; Donald A. Chant to Boyce, 24 November 1965, Fleschner file, UCR-DE; DeBach to Boyce, 1 February 1961, PHD; Evert I. Schlinger to DeBach, 27 August 1962, PHD; Steinhaus, *Disease in a Minor Chord,* pp. 235–240. Steinhaus's department was named Insect Pathology, not Invertebrate Pathology as he desired.

84. Boyce, DeBach, Huffaker, Kennett, interviews; Schlinger to DeBach, 12 June 1963, PHD; Paul DeBach, "Preface," in DeBach, ed., *Biological Control of Insect Pests and Weeds,* pp. v–vi. Disputes over who would contribute to the book began as early as 1953, when Steinhaus argued that his associates in insect pathology should be allowed to join him on that section. The outline as of 1957 included Smith, Clausen, Compere, Flanders, and Fleschner, all of whom eventually dropped out. Steinhaus to Clausen 24 November 1953, EAS; Outline for biological control book, 1957, PHD.

85. Van den Bosch, *Pesticide Conspiracy,* p. 199.

86. Boyce, Fleschner, Oatman, Doutt, interviews; interview with Irvin M. Hall, 30 July 1987; DeBach to Boyce, 1 February 1961, PHD; Boyce to Bartlett, 7 April 1961, Fleschner file, UCR-DE.

87. Steinhaus, *Disease in a Minor Chord,* pp. 243–247; Boyce, Doutt, Fisher, Hall, Oatman, interviews. Steinhaus stayed in the entomology department only a few months. Aldrich was named chancellor of the new Irvine campus, and he made Steinhaus dean of biological sciences.

88. Aldrich to Fleschner, 16 February 1962, Fleschner file, UCR-DE; Albert L. Turnbull and Donald A. Chant, "The practice and theory of biological control of insects in Canada," *Canadian Journal of Zoology* 39 (1961): 697–753; Donald A. Chant, "Strategy and tactics of insect control," *Canadian Entomologist* 96 (1964): 182–201; idem, "Integrated control systems," in *Scientific Aspects of Pest Control* (Washington: National Academy of Sciences and National Research Council, 1966), pp. 193–218.

89. Huffaker and Kennett, "Biological control of *Parlatoria oleae*"; Carl B. Huffaker, F. J. Simmonds, and J. E. Laing, "The theoretical and empirical basis of biological control," in *Theory and Practice of Biological Control,* ed. Carl B. Huffaker and Powers S. Messenger (New York: Academic Press, 1976), pp. 41–78; Harold Compere, "Changing trends and objectives in biological control," *Proceedings of the First International Citrus Symposium* 2 (1969): 755–764.

90. Palladino, "Entomology and ecology," pp. 112–138, 253–255; idem, "Ecological theory and pest control practice: A study of the institutional and conceptual dimensions of a scientific debate," *Social Studies of Science* 20 (1990): 255–281.

91. DeBach, interview.

92. Hall, interview.

93. Tom Patterson, "Chemical, biological advocates join forces at UCR," *Riverside Daily Enterprise,* 31 July 1969; Hall, interview; interview with Ivan Hinderaker, 30 June 1988; interview with W. Mack Dugger, 30 June 1988. Boyce, a firm believer in the mission-oriented research outlook on which the Citrus Experiment Station was founded, vehemently opposed Hinderaker's actions bringing agriculture and the other scientific departments together. Boyce believed that Hinderaker built up the rest of the campus at the expense of agriculture. Boyce, *Odyssey of an Entomologist,* pp. 288–289; idem, interview.

Chapter 7. CONCLUSION: AGRIBUSINESS AND BIOLOGICAL CONTROL

1. Jim Hightower, *Hard Tomatoes, Hard Times* (Cambridge, Mass.: Shenkman, 1973); John H. Perkins, *Insects, Experts, and the Insecticide Crisis: The Quest for New Pest Management Strategies* New York: Plenum, 1981); idem, "Insects, food, and hunger: The paradox of plenty for U.S. entomology," *Environmental Review* 7 (1983): 71–86; Robert van den Bosch, *The Pesticide Conspiracy* (Garden City, N.Y.: Doubleday, 1978).

2. For examples, see Louis A. McLean, "Pesticides and the environment," *Bioscience* 17 (1967): 613–617; and Jamie L. Whitten, *That We May Live* (Princeton, N.J.: Van Nostrand, 1966), pp. 132–138.

3. John H. Perkins, "The quest for innovation in agricultural entomology, 1945–1978," in *Pest Control: Cultural and Environmental Aspects,* ed. David Pimentel and John H. Perkins (Boulder, Colo.: Westview Press, 1980), pp. 23–80, on pp. 63–64.

4. Howard B. Lorbeer, "The organization and operation of a pest control district," *California Citrograph* 16 (1931): 451, 484; "Putting bugs to work" (editorial), ibid., 33 (1948): 377; Paul DeBach, *Biological Control by Natural Enemies* (London: Cambridge University Press, 1974), pp. 286–288; Robert van den Bosch, Powers S. Messenger, and A. P. Gutierrez, *An Introduction to Biological Control* (New York: Plenum, 1982), pp. 216–218.

5. Alfred M. Boyce, *Odyssey of an Entomologist: Adventures on the Farm, at Sea, and in the University* (Riverside, Calif.: UC Riverside Foundation, 1987), pp. 144–145; interview with Daniel G. Aldrich, Jr., 1 July 1988; interview with Alfred M. Boyce, 13–14 May 1987; interview with Earl R. Oatman, 24 July 1987; interview with Ivan Hinderaker, 30 June 1988; interview with Charles E. Kennett, 17 June 1987; interview with Ray F. Smith, 13 June 1987; interview with Carl B. Huffaker, 10 June 1987.

6. Frank Graham, Jr., *The Dragon Hunters* (New York: E. P. Dutton, 1984), p. 16; Whitten, *That We May Live,* pp. 184–185.

7. Francis G. Howarth, "Classical biocontrol: Panacea or Pandora's box," *Proceedings of the Hawaiian Entomological Society* 24 (1983): 239–244; idem, "Environmental impacts of classical biological control," *Annual Review of Entomology* 36 (1991): 485–509.

8. Jack Ralph Kloppenburg, Jr., *First the Seed: The Political Economy of Plant Biotechnology, 1492–2000* (Cambridge: Cambridge University Press, 1988); Jack Doyle, *Altered Harvest: Agriculture, Genetics, and the Fate of the World's Food Supply* (New York: Viking, 1985).

9. Anne Simon Moffat, "Research on biological pest control moves ahead," *Science* 252 (1991): 211–212.

10. Richard Garcia, Leopoldo E. Caltagirone, and Andrew P. Gutierrez, "Comments on a redefinition of biological control," *Bioscience* 38 (1988): 692–694.

11. Foreword to Robert van den Bosch, *The Pesticide Conspiracy* (Berkeley: University of California Press, 1989), pp. vii–xiv, quotation from p. xiii.

12. Paolo S. A. Palladino, "Entomology and ecology: The ecology of entomology" (Ph.D. dissertation, University of Minnesota, 1989), pp. 243–255.

13. Ibid., p. 275.

14. John Harley Warner, "Physiological theory and therapeutic explanation in the 1860s: The British debate on the medical use of alcohol," *Bulletin of the History of Medicine* 54 (1980): 235–257, on p. 236.

15. Paul DeBach, "Successes, trends, and future possibilities," in *Biological Control of Insect Pests and Weeds,* ed. Paul DeBach (New York: Reinhold, 1964), pp. 673–713; Albert L. Turnbull and Donald A. Chant, "The practice and theory of biological control of insects in Canada," *Canadian Journal of Zoology* 39 (1961): 697–753.

16. John M. Staudenmaier, S.J., *Technology's Storytellers: Reweaving the Human Fabric* (Cambridge, Mass.: Society for the History of Technology and MIT Press, 1985), pp. 85-103; Edwin T. Layton, Jr., "Through the looking glass, or news from Lake Mirror Image," *Technology and Culture* 28 (1987): 594-607, on pp. 599-600.

17. Deborah Fitzgerald, *The Business of Breeding: Hybrid Corn in Illinois, 1890-1940* (Ithaca, N.Y.: Cornell University Press, 1990), p. 1.

18. Pamela M. Henson, "Evolution and taxonomy: J. H. Comstock's research school in evolutionary entomology at Cornell University, 1874-1930" (Ph.D. dissertation, University of Maryland, College Park, 1990), pp. 350-351.

19. Van den Bosch, *Pesticide Conspiracy,* pp. 47-54, 99-101.

20. Peter J. Bowler, "Science and the environment: New agendas for the history of science?" in *Science and Nature: Essays in the History of the Environmental Sciences,* ed. Michael Shortland ([Faringdon, Engl.]: British Society for the History of Science, 1993), pp. 1-21.

21. Perkins, *Insects, Experts, and the Insecticide Crisis*; Paolo Palladino, "On 'environmentalism': The origins of debates over policy for pest-control research in America, 1960-1975," in Shortland, ed., *Science and Nature,* pp. 181-212.

22. Richard L. Doutt, "Biological control: The California connection," keynote address, International Vedalia Symposium on Biological Control: A Century of Success, Riverside, Calif., 27 March 1989. I was privileged to attend.

Bibliography

"Agricultural patrols by airplane." 1919. *California Department of Agriculture Bulletin* 8: 101–103.

Ainsworth, Ed. 1968. *Journey with the Sun: The Story of Citrus in its Western Pilgrimage.* Los Angeles: Sunkist Growers.

Anderson, Oscar E., Jr. 1953. *Refrigeration in America: A History of a New Technology and its Impact.* Princeton: Princeton University Press.

Andrewartha, Herbert G., and Birch, L. Charles. 1954. *The Distribution and Abundance of Animals.* Chicago: University of Chicago Press.

_____. 1960. "Some recent contributions to the study of the distribution and abundance of insects." *Annual Review of Entomology* 5: 219–242.

Die Angewandte Entomologie in den Vereinigten Staaten, by Karl Escherich, review. 1913. *Journal of Economic Entomology* 6: 430–432.

Armitage, Horace Morton. 1919. "Controlling mealybugs by the use of their natural enemies." *California State Commission of Horticulture Bulletin* 8: 257–260.

_____. 1920. "Report of the biological control work directed against the mealybugs." *California Department of Agriculture Bulletin* 9: 441–451.

_____. 1924. "The citrophilus mealybug, *Pseudococcus gahani* Green, as a major pest of citrus in southern California." *Journal of Economic Entomology* 17: 554–561.

_____. 1929. "Timing field liberations of Cryptolaemus in the control of citrophilus mealybug in the infested citrus orchards of southern California." *Journal of Economic Entomology* 22: 910–915.

_____. 1949. "The oriental fruit fly from the mainland viewpoint." *Journal of Economic Entomology* 42: 713–716.

Armstrong, Paul S. 1923. "Sunkist advertising—how it sells California oranges and lemons." *California Citrograph* 8: 222–223, 238–239.

"Australian parasites of the fluted scale." 1888. *Pacific Rural Press* 35: 345.

Baker, W. A.; Bradley, W. G.; and Clark, Charles A. 1949. "Biological control of the European corn borer in the United States." *U.S. Department of Agriculture Technical Bulletin* 983.

Barnes, Barry. 1982. "The science-technology relationship: A model and a query." *Social Studies of Science* 12: 166–172.

Bartlett, Blair R. 1953. "Retentive toxicity of field-weathered insecticide residues to entomophagous insects associated with citrus in California." *Journal of Economic Entomology* 46: 565–569.

_____. 1956. "Natural predators: Can selective insecticides help to preserve biotic control?" *Agricultural Chemicals* 11, no. 2: 42–44, 107–109.

_____. 1957. "Where are we going in biological control?" *California Citrograph* 42: 339, 360–361.

_____. 1962. "The international shipment of adult entomophagous insects." *Annals of the Entomological Society of America* 55: 448–455.

Bateson, William. 1913. *Problems of Genetics*. New Haven: Yale University Press.

Beechert, Edward D. 1985. *Working in Hawaii: A Labor History*. Honolulu: University of Hawaii Press.

Belknap, Michael R. 1968. "The era of the lemon: A history of Santa Paula, California." *California Historical Society Quarterly* 47: 113–140.

"Biological control of red and black scale discussed by Professor Smith." 1937. *California Citrograph* 23: 59, 87.

Bodenheimer, Friedrich Simon. 1928. "Welche Faktoren regulieren die Individuenzahl einer Insektenart in der Natur?" *Biologisches Zentralblatt* 48: 714–739.

_____. 1930. "Über die Grundlagen einer allgemeinen Epidemiologie der Insektenkalamitäten." *Zeitschrift für Angewandte Entomologie* 16: 433–450.

_____. 1938. *Problems of Animal Ecology*. London: Oxford University Press.

_____. 1959. *A Biologist in Israel*. Jerusalem: Turin Press.

Bowler, Peter J. 1983. *The Eclipse of Darwinism: Anti-Darwinian Evolution Theories in the Decades around 1900*. Baltimore: Johns Hopkins University Press.

Boyce, Alfred M. 1950. "Entomology of citrus and its contribution to entomological principles and practices." *Journal of Economic Entomology* 43: 741–766.

_____. 1987. *Odyssey of an Entomologist: Adventures on the Farm, at Sea, and in the University*. Riverside, Calif.: UC Riverside Foundation.

Boyce, Alfred M., and Ewart, William H. 1946. "DDT for control of citrus thrips and citricola scale in central California." *California Citrograph* 31: 240–241.

Boyce, Robert. 1987. "Insects and international relations: Canada, France, and British agricutural sanitary import restrictions between the Wars." *International History Review* 9: 1–27.

Branigan, E. J. 1915. "Vedalia vs. icerya on pears." *California State Commission of Horticulture Bulletin* 4: 107–108.

_____. 1916. "A satisfactory method of rearing mealybugs for use in parasite work." *California State Commission of Horticulture Bulletin* 5: 304–306.

"Bread on the waters." 1934. *California Citrograph* 19: 261.

Broodryk, S. W., and Doutt, Richard L. 1966. "The biology of *Coccophagoides utilis* Doutt (Hymenoptera, Aphelinidae)." *Hilgardia* 37: 233–254.

Caltagirone, Leopoldo E., and Doutt, Richard L. 1989. "The history of the vedalia beetle importation to California and its impact on the development of biological control." *Annual Review of Entomology* 34: 1–16.

Carnes, Edward K. 1912. "Collecting ladybirds (Coccinellidae) by the ton." *California State Commission of Horticulture Bulletin* 1: 71–81.

_____. 1912. "Insectary Division. Report for the month of May." *California State Commission of Horticulture Bulletin* 1: 820–828.

Carson, Rachel. 1962. *Silent Spring.* Boston: Houghton Mifflin.

Chamberlin, Thomas R. 1926. "The introduction and establishment of the alfalfa weevil parasite, *Bathyplectes curculionis* (Thoms.), in the United States." *Journal of Economic Entomology* 19: 302–310.

Chambers, Clarke A. 1952. *California Farm Organizations: A Historical Study of the Grange, the Farm Bureau, and the Associated Farmers.* Berkeley: University of California Press.

Chant, Donald A. 1964. "Strategy and tactics of insect control." *Canadian Entomologist* 96: 182–201.

_____. 1966. "Integrated control systems." In *Scientific Aspects of Pest Control,* pp. 193–218. Washington: National Academy of Sciences and National Research Council.

Chapman, Royal N. 1929. "The potentialities of entomology." *Science* 69: 413–418.

_____. 1931. *Animal Ecology with Especial Reference to Insects.* New York: McGraw-Hill.

Clancy, Donald W., and Muma, Martin H. 1959. "Purple scale parasite found in Florida." *Journal of Economic Entomology* 52: 1025–1026.

Clausen, Curtis P. 1915. "Mealy bugs of citrus trees." *California Agricultural Experiment Station Bulletin* 258: 17–48.

_____. 1936. "Insect parasitism and biological control." *Annals of the Entomological Society of America* 29: 201–223.

_____. 1942. "The relation of taxonomy to biological control." *Journal of Economic Entomology* 35: 744–748.

_____. 1956. "Biological control of insect pests in the continental United States." *U.S. Department of Agriculture Technical Bulletin* 1139.

_____. 1978. "Biological control of citrus insects." In *The Citrus Industry,* vol. 4: *Crop Protection,* edited by Walter Reuther, E. Clair Calavan, and Glenn E. Carman, pp. 276–320. Berkeley: University of California Division of Agricultural Sciences.

_____, ed. 1978. *Introduced Parasites and Predators of Arthropod Pests and Weeds: A World Review.* Agriculture Handbook 480. Washington: U.S. Department of Agriculture.

Clausen, Curtis P., and Berry, Paul A. 1932. "The citrus blackfly in Asia, and the importation of its natural enemies into tropical America." *U.S. Department of Agriculture Technical Bulletin* 320.

Clausen, Curtis P.; Clancy, Donald W.; and Chock, Q. C. 1965. "Biological control of the oriental fruit fly (*Dacus dorsalis* Hendel) and other fruit flies in Hawaii." *U.S. Department of Agriculture Technical Bulletin* 1322.

Clements, Frederic E., and Shelford, Victor E. 1939. *Bio-Ecology.* New York: John Wiley & Sons.

Coit, J. Eliot. 1915. *Citrus Fruits: An Account of the Citrus Fruit Industry with Special Reference to California Requirements and Practices and Similar Conditions.* New York: Macmillan.

Compere, George. 1912. "A few facts concerning the fruit flies of the world." *California State Commission of Horticulture Bulletin* 1: 709–730, 842–845, 907–911, 929–932.

_____. 1922. "Origin of fumigation with hydrocyanic acid gas in California." *California Department of Agriculture Bulletin* 11: 438–442.

Compere, Harold. 1922. "The black scale problem." *California Cultivator* 59: 29–30.

_____. 1928. "Successful importation of five new natural enemies of citrophilus mealybug from Australia." *California Citrograph* 13: 318, 346–349.

_____. 1930. "Hazards of parasite hunting in foreign lands." *California Citrograph* 15: 533, 562–568.

_____. 1935. "The search for parasites of red scale and other citrus pests in Brazil." *Citrus Leaves* 15, no. 1: 8–9.

_____. 1937. "Collecting red and black scale parasites in Africa." *California Citrograph* 23: 58, 88–89.

_____. 1939. "The insect enemies of the black scale, *Saissetia oleae* (Bern.), in South America." *University of California Publications in Entomology* 7: 75–90.

_____. 1940. "Parasites of the black scale, *Saissetia oleae,* in Africa." *Hilgardia* 13: 387–425.

_____. 1953. "An appraisal of Silvestri's work in the Orient for the University of California, some misidentifications corrected, and two forms of *Casca* described as new species." *Bollettino del Laboratorio di zoologia generale e agraria della Facolta agraria in Portici* 33: 35–46.

_____. 1955. "A systematic study of the genus *Aphytis* Howard (Hymenoptera, Aphelinidae) with descriptions of new species." *University of California Publications in Entomology* 10: 271–320.

_____. 1961. "The red scale and its insect enemies." *Hilgardia* 31: 173–278.

_____. 1969. "The role of systematics in biological control: A backward look." *Israel Journal of Entomology* 4: 5–10.

_____. 1969. "Changing trends and objectives in biological control." *Proceedings of the First International Citrus Symposium* 2: 755–764.

_____. 1971. "Your health—my oranges." *Citrograph* 55: 324–327.

Compere, Harold; Flanders, Stanley E.; and Smith, Harry S. 1941. "Use air transport from China for the introduction into California of a red scale inhabiting *Comperiella.*" *California Citrograph* 26: 291, 300–301.

Compere, Harold, and Smith, Harry S. 1927. "Notes on the life-history of two oriental chalcidoid parasites of *Chrysomphalus.*" *University of California Publications in Entomology* 4: 63–73.

_____. 1932. "The control of the citrophilus mealybug, *Pseudococcus gahani,* by Australian parasites." *Hilgardia* 6: 585–618.

Cook, Albert John. 1911. "The State Horticultural Commission." *California State Commission of Horticulture Bulletin* 1: 30–34.

Cooper, Ellwood. 1913. *Bug vs. Bug: Parasitology.* Santa Barbara, Calif.

Coppel, Harry C., and Mertins, James W. 1977. *Biological Insect Pest Suppression.* Berliner: Springer-Verlag.

Craighead, Frank C. 1921. "Hopkins host-selection principle as related to certain cerambycid beetles." *Journal of Agricultural Research* 22: 189–220.

_____. 1923. "The host-selection principle as advanced by Walsh." *Canadian Entomologist* 55: 76–80.

Craw, Alexander. 1902. "Horticultural quarantine reports." *Biennial Report of the California State Board of Horticulture* 1901–02: 185–204.

Cumberland, William W. 1917. *Cooperative Marketing: Its Advantages as Exemplified in the California Fruit Growers Exchange.* Princeton: Princeton University Press.

Curtis, John. 1860. *Farm Insects.* Glasgow: Blackie and Son.

"Damaging facts *in re* Board of Horticulture." 1895. *Rural Californian,* March 1895, pp. 144–146.

Danbom, David B. 1986. "The agricultural experiment station and professionalization: Scientists' goals for agriculture." *Agricultural History* 60, no. 2: 246–255.

Daniel, Pete. 1985. *Breaking the Land: The Transformation of Cotton, Tobacco, and Rice Cultures Since 1880.* Urbana: University of Illinois Press, 1985.

Darwin, Charles. 1859. *On the Origin of Species by Means of Natural Selection.* London: John Murray.

Davidson, W. M. 1919. "The convergent ladybeetle (*Hippodamia convergens* Guerin) and the barley-corn aphis (*Aphis maidis* Fitch)." *California State Commission of Horticulture Bulletin* 8: 23–26.

_____. 1925. "Observations and experiments on the dispersion of the convergent lady-beetle (*Hippodamia convergens* Guérin) in California." *Transactions of the American Entomological Society* 50: 163–175.

Davis, C. J., and Krauss, Noel L. H. 1962. "Recent developments in the biological control of weed pests in Hawaii." *Proceedings of the Hawaiian Entomological Society* 18: 65–67.

DeBach, Paul. 1946. "An insecticidal check method for measuring the efficacy of entomophagous insects." *Journal of Economic Entomology* 39: 695–697.

_____. 1947. "Cottony cushion scale, vedalia, and DDT in central California." *California Citrograph* 32: 406–407.

_____. 1951. "The necessity for an ecological approach to pest control in citrus in California." *Journal of Economic Entomology* 44: 443–447.

_____. 1952. "Biological control of red scale in San Diego County." *California Citrograph* 37: 136–137, 158–160.

_____. 1953. "Purple scale parasites in southern California." *California Citrograph* 38: 219–222.

_____. 1958. "Selective breeding to improve adaptations of parasitic insects." *Proceedings of the Tenth International Congress of Entomology* 4: 759–768.

_____. 1960. "The importance of taxonomy to biological control as illustrated by the cryptic history of *Aphytis holoxanthus* n. sp. (Hymenoptera: Aphelinidae), a parasite of *Chrysomphalus aonidum,* and *Aphytis coheni* n sp., a parasite of *Aonidiella aurantii.*" *Annals of the Entomological Society of America* 53: 701–705.

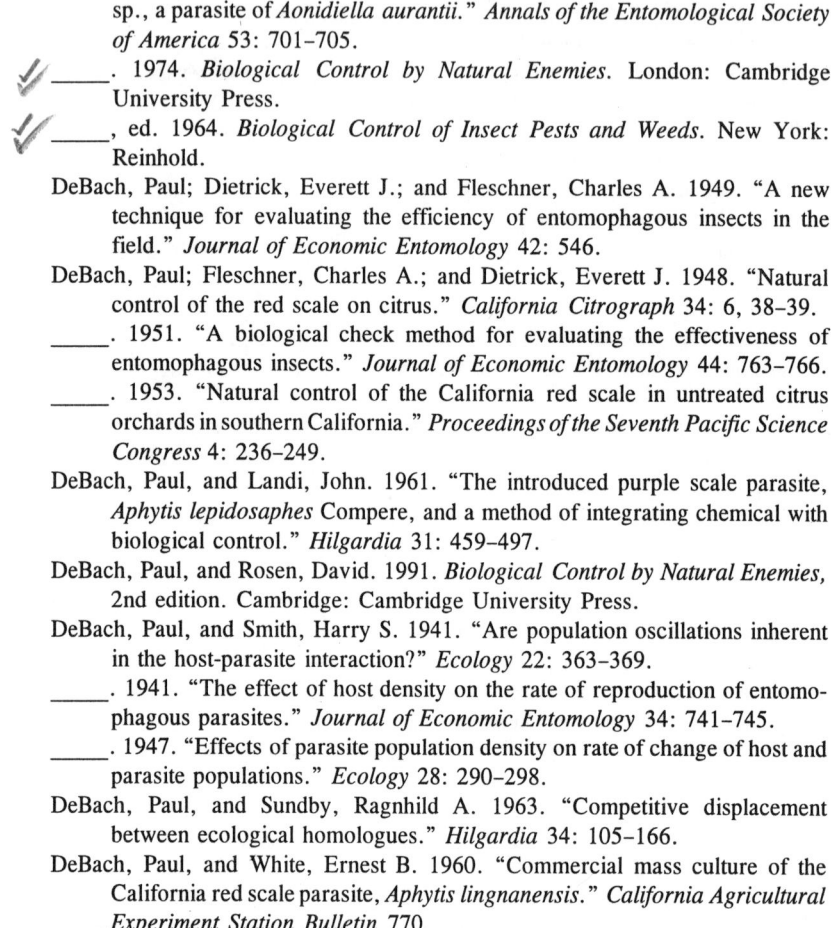

_____. 1974. *Biological Control by Natural Enemies.* London: Cambridge University Press.

_____, ed. 1964. *Biological Control of Insect Pests and Weeds.* New York: Reinhold.

DeBach, Paul; Dietrick, Everett J.; and Fleschner, Charles A. 1949. "A new technique for evaluating the efficiency of entomophagous insects in the field." *Journal of Economic Entomology* 42: 546.

DeBach, Paul; Fleschner, Charles A.; and Dietrick, Everett J. 1948. "Natural control of the red scale on citrus." *California Citrograph* 34: 6, 38–39.

_____. 1951. "A biological check method for evaluating the effectiveness of entomophagous insects." *Journal of Economic Entomology* 44: 763–766.

_____. 1953. "Natural control of the California red scale in untreated citrus orchards in southern California." *Proceedings of the Seventh Pacific Science Congress* 4: 236–249.

DeBach, Paul, and Landi, John. 1961. "The introduced purple scale parasite, *Aphytis lepidosaphes* Compere, and a method of integrating chemical with biological control." *Hilgardia* 31: 459–497.

DeBach, Paul, and Rosen, David. 1991. *Biological Control by Natural Enemies,* 2nd edition. Cambridge: Cambridge University Press.

DeBach, Paul, and Smith, Harry S. 1941. "Are population oscillations inherent in the host-parasite interaction?" *Ecology* 22: 363–369.

_____. 1941. "The effect of host density on the rate of reproduction of entomophagous parasites." *Journal of Economic Entomology* 34: 741–745.

_____. 1947. "Effects of parasite population density on rate of change of host and parasite populations." *Ecology* 28: 290–298.

DeBach, Paul, and Sundby, Ragnhild A. 1963. "Competitive displacement between ecological homologues." *Hilgardia* 34: 105–166.

DeBach, Paul, and White, Ernest B. 1960. "Commercial mass culture of the California red scale parasite, *Aphytis lingnanensis.*" *California Agricultural Experiment Station Bulletin* 770.

DeLotto, G. 1958. "The Pseudococcidae (Hom.: Coccoidea) described by C. K. Brain from South Africa." *Bulletin of the British Museum (Natural History), Entomology* 7: 79–120.

Dickson, Robert C. 1941. "Inheritance of resistance to hydrocyanic acid fumigation in the California red scale." *Hilgardia* 13: 513–521.

_____. 1949. "Factors governing the induction of diapause in the oriental fruit moth." *Annals of the Entomological Society of America* 42: 511–537.

Doane, Charles C., and McManus, Michael L., eds. 1981. "The gypsy moth:

Research toward integrated pest management." *U.S. Department of Agriculture Technical Bulletin* 1584.

Dobzhansky, Theodosius. 1935. "A critique of the species concept in biology." *Philosophy of Science* 2: 344–355.

_____. 1935. "*Drosophila miranda,* a new species." *Genetics* 20: 377–391.

_____. 1937. *Genetics and the Origin of Species.* New York: Columbia University Press.

_____. 1941. *Genetics and the Origin of Species,* 2nd edition. New York: Columbia University Press.

Dodd, Alan P. 1940. *The Biological Campaign Against Prickly Pear.* Brisbane: Commonwealth Prickly Pear Board.

Doutt, Richard L. 1954. "An evaluation of some natural enemies of the olive scale." *Journal of Economic Entomology* 47: 39–43.

_____. 1958. "Vice, virtue, and the vedalia." *Bulletin of the Entomological Society of America* 4: 119–123.

_____. 1959. "The biology of parasitic Hymenoptera." *Annual Review of Entomology* 4: 161–182.

_____. 1966. "A taxonomic analysis of parasitic Hymenoptera reared from *Parlatoria oleae* (Colvée)." *Hilgardia* 37: 219–231.

Doyle, Jack. 1985. *Altered Harvest: Agriculture, Genetics, and the Fate of the World's Food Supply.* New York: Viking.

Dunlap, Thomas R. 1978. "The triumph of chemical pesticides in insect control 1890–1920." *Environmental Review* 1, no. 5: 38–47.

_____. 1978. "Science as a guide in regulating technology: The case of DDT in the United States." *Social Studies of Science* 8: 265–285.

_____. 1981. *DDT: Scientists, Citizens, and Public Policy.* Princeton: Princeton University Press.

Dupree, A. Hunter. 1957. *Science in the Federal Government: A History of Policies and Activities to 1940.* Cambridge, Mass.: Harvard University Press, Belknap Press.

Ebeling, Walter. 1950. *Subtropical Entomology.* San Francisco: Lithotype Process Co.

Eichner, Alfred S. 1969. *The Emergence of Oligopoly: Sugar Refining as a Case Study.* Baltimore: Johns Hopkins University Press.

Elton, Charles S. 1958. *The Ecology of Invasions by Animals and Plants.* London: Methuen.

Emerson, Alfred E. 1935. "Termitophile distribution and quantitative characters as indicators of physiological speciation in British Guiana termites." *Annals of the Entomological Society of America* 28: 369–395.

"Entomological explorer returns." 1948. *California Citrograph* 34: 70–71.

"Entomology notes." 1970. *Citrograph* 55: 184.

Erdman, Henry E. 1958. "The development and significance of California cooperatives." *Agricultural History* 32: 179–184.

Escherich, Karl. 1913. *Die Angewandte Entomologie in den Vereinigten Staaten.* Berlin: Paul Parey.

Essig, Edward O. 1931. *A History of Entomology*. New York: Macmillan.

———. 1942. "The significance of taxonomy in the general field of economic entomoloy." *Journal of Economic Entomology* 35: 739–743.

———. 1955. "Official entomology in California—some comments, historical and personal." *California Department of Agriculture Bulletin* 44: 3–16.

Essig, Edward O., and Michelbacher, Abraham E. 1933. "The alfalfa weevil." *California Agricultural Experiment Station Bulletin* 567.

Evans, Hughes. 1989. "European malaria policy in the 1920s and 1930s: The epidemiology of minutiae." *Isis* 80: 40–59.

Ferris, Gordon F. 1919. "Observations on some mealy-bugs (Hemiptera; Coccidae)." *Journal of Economic Entomology* 12: 292–299.

———. 1927. "Mealybugs." *California Department of Agriculture Bulletin* 16: 336–342.

———. 1928. "The principles of systematic entomology." *Stanford University Publications, University Series, Biological Sciences* 5: 101–270.

———. 1930. "The effectiveness of a plant quarantine." *Science* 71: 68–69.

———. 1930. "The plant quarantines once more." *Science* 71: 606–607.

———. 1942. "The needs of systematic entomology." *Journal of Economic Entomology* 35: 732–738.

Finney, Glenn L.; Flanders, Stanley E.; and Smith, Harry S. 1947. "Mass culture of *Macrocentrus ancylivorus* and its host, the potato tuber moth." *Hilgardia* 17: 437–483.

Fiske, William F. 1910. "Superparasitism: An important factor in the natural control of insects." *Journal of Economic Entomology* 3: 88–97.

———. 1913. "The gipsy moth as a forest insect, with suggestions as to its control." *U.S. Bureau of Entomology Circular* 164.

Fitch, Asa. 1860. "Address, on our most pernicious insects." *Transactions of the New York State Agricultural Society* 20: 588–598.

———. 1860. "Sixth report on the noxious and other insects of the state of New York." *Transactions of the New York State Agricultural Society* 20: 745–868.

Fitzgerald, Deborah. 1990. *The Business of Breeding: Hybrid Corn in Illinois, 1890–1940*. Ithaca, N.Y.: Cornell University Press.

Flanders, Stanley E. 1925. "A new departure in codling moth control." *Journal of Economic Entomology* 18: 838–839.

———. 1929. "The mass production of *Trichogramma minutum* Riley and observations on the natural and artificial parasitism of the codling moth egg." *Transactions of the Fourth International Congress of Entomology* 2: 110–130.

———. 1930. "Mass production of egg parasites of the genus *Trichogramma*." *Hilgardia* 4: 465–501.

———. 1930. "Recent developments in *Trichogramma* production." *Journal of Economic Entomology* 23: 837–841.

———. 1931. "The temperature relationships of *Trichogramma minutum* as a basis for racial segregation." *Hilgardia* 5: 395–406.

_____. 1932. "Observations and experiences in searching for parasites in Australia." *Citrus Leaves* 12, no. 4: 15–18.

_____. 1936. "A biological phenomenon affecting the establishment of Aphelinidae as parasites." *Annals of the Entomological Society of America* 29: 251–255.

_____. 1937. "Habitat selection by *Trichogramma*." *Annals of the Entomological Society of America* 30: 208–210.

_____. 1937. "Ovipositional instincts and developmental sex differences in the genus *Coccophagus*." *University of California Publications in Entomology* 6: 401–422.

_____. 1938. "Identity of the common species of American *Trichogramma*." *Journal of Economic Entomology* 31: 456–457.

_____. 1939. "The practical application of biological studies of parasites employed in biological control." *Proceedings of the Sixth Pacific Science Congress* 4: 373–381.

_____. 1942. "*Metaphycus helvolus,* an encyrtid parasite of the black scale." *Journal of Economic Entomology* 35: 690–698.

_____. 1949. "George Compere, pioneer in the biological control of red scale." *California Citrograph* 34: 160–162.

_____. 1951. "Mass culture of California red scale and its golden chalcid parasites." *Hilgardia* 21: 1–42.

_____. 1953. "Hymenopterous parasites of three species of oriental scale insects." *Bollettino del Laboratorio di zoologia generale e agraria della Facolta agraria in Portici* 33: 10–28.

_____. 1954. "*Casca*'s elusive husband." *California Citrograph* 39: 343, 352.

_____. 1955. "The organization of biological control and its historical development." *Mededelingen van de Landbouwhogeschool en de Opzoekingsstations van de Staat te Gent* 20: 257–270.

_____. 1957. "Fig scale parasites introduced into California." *Journal of Economic Entomology* 50: 171–172.

_____. 1964. "Some biological control aspects of taxonomy exemplified by the genus *Aphytis* (Hymenoptera: Aphelinidae)." *Canadian Entomologist* 96: 888–893.

_____. 1965. "Competition and cooperation among parasitic Hymenoptera related to biological control." *Canadian Entomologist* 97: 409–422.

Flanders, Stanley E.; Gressitt, J. Linsley; and Fisher, Theodore W. 1958. "*Casca chinensis,* an internal parasite of California red scale." *Hilgardia* 28: 65–91.

Fleming, Walter E. 1968. "Biological control of the Japanese beetle." *U.S. Department of Agriculture Technical Bulletin* 1383.

Fleschner, Charles A. 1958. "Field approach to population studies of tetranychid mites on citrus and avocado in California." *Proceedings of the Tenth International Congress of Entomology* 2: 669–674.

_____. 1958. "Natural enemies of tetranychid mites on citrus and avocado in southern California." *Proceedings of the Tenth International Congress of*

Entomology 4: 627–631.

"The fluted scale." 1888. *Pacific Rural Press* 35: 110.

"Follow up the advantage." 1931. *California Citrograph* 16: 449.

Frison, Theodore H. 1942. "The significance of economic entomology in the field of insect taxonomy." *Journal of Economic Entomology* 35: 749–752.

Galloway, J. H. 1989. *The Sugar Cane Industry: An Historical Geography from its Origins to 1914.* Cambridge: Cambridge University Press.

Garcia, Richard; Caltagirone, Leopoldo E.;, and Gutierrez, Andrew P. 1988. "Comments on a redefinition of biological control." *Bioscience* 38: 692–694.

"Get that beetle." 1937. *California Citrograph* 22: 233.

Giffard, Walter M. 1920. "A review of the organization of the Hawaiian Entomological Society and brief mention of some of the more notable achievements in Hawaii by its members." *Proceedings of the Hawaiian Entomological Society* 4: 363–373.

Goeden, Richard D.; Fleschner, Charles A.; and Ricker, Donald W. 1967. "Biological control of prickly pear cacti on Santa Cruz Island, California." *Hilgardia* 38: 579–606.

Graham, Frank, Jr. 1984. *The Dragon Hunters.* New York: E. P. Dutton.

Greene, John C. 1959. *The Death of Adam: Evolution and its Impact on Western Thought.* Ames: Iowa State University Press.

Gressitt, J. Linsley, and Flanders, Stanley E. 1949. "New developments in the transport of beneficial insects." *Journal of Economic Entomology* 42: 150.

Griffiths, James T. 1951. "Possibilities for better citrus insect control through the study of the ecological effects of spray programs." *Journal of Economic Entomology* 44: 464–468.

_____. 1953. "Where do we go from here?" *Florida Entomologist* 36: 135–140.

Griffiths, James T.; Stearns, Charles R., Jr.; and Thompson, William L. 1951. "Parathion hazards encountered spraying citrus in Florida." *Journal of Economic Entomology* 44: 160–163.

Griffiths, James T., and Thompson, William L. 1953. "Reduced spray programs for citrus for canning plants in Florida." *Journal of Economic Entomology* 46: 930–936.

Gunther, Francis A., and Jeppson, Lee R. 1960. *Modern Insecticides and World Food Production.* London: Chapman and Hall.

Hafez, Mostafa, and Doutt, Richard L. 1954. "Biological evidence of sibling species in *Aphytis maculicornis* (Masi) (Hymenoptera, Aphelinidae)." *Canadian Entomologist* 86: 90–96.

Hagen, Joel B. 1988. "Organism and environment: Frederic Clements's version of a unified physiological ecology." In *The American Development of Biology,* edited by Ronald Rainger, Keith R. Benson, and Jane Maienschein, pp. 257–280. Philadelphia: University of Pennsylvania Press.

_____. 1992. *An Entangled Bank: The Origins of Ecosystem Ecology.* New Brunswick: Rutgers University Press.

"Harry S. Smith retires." 1951. *California Citrograph* 36: 380–382.

Hatch, Melville H., and Tanasse, Cherie. 1948. "The liberation of *Hippodamia convergens* in the Yakima valley of Washington, 1943 to 1946." *Journal of Economic Entomology* 41: 993.

Hecke, George H. 1919. "Ellwood Cooper." *California State Commission of Horticulture Bulletin* 8: inside title page.

Helms, Douglas. 1979. "Technological methods for boll weevil control." *Agricultural History* 53: 286–299.

Henson, Pamela M. 1990. "Evolution and taxonomy: J. H. Comstock's research school in evolutionary entomology at Cornell University, 1874–1930." Ph.D. dissertation, University of Maryland, College Park.

Herms, William B. 1926. "An analysis of some of California's major entomological problems." *Journal of Economic Entomology* 19: 262–270.

Heyboer, Maarten. 1992. "Grass-counters, stock-feeders, and the dual orientation of applied science: The history of range science, 1895–1960." Ph.D. dissertation, Virginia Polytechnic Institute and State University.

Hightower, Jim. 1973. *Hard Tomatoes, Hard Times*. Cambridge, Mass.: Shenkman.

Hinds, Warren E.; Osterberger, B. A.; and Dugas, A. L. 1933. "Sugar cane borer control by *Trichogramma* colonization in Louisiana in 1932." *Journal of Economic Entomology* 26: 758–767.

"H. M. Armitage." 1931. *California Department of Agriculture Bulletin* 20: 598.

Holloway, James K., and Huffaker, Carl B. 1953. "Insects to control a weed." *U.S. Department of Agriculture Yearbook* 1952: 135–140.

Hopkins, James T. 1960. *Fifty Years of Citrus: The Florida Citrus Exchange, 1909–1959*. Gainesville: University of Florida Press.

Hoskins, William M.; Borden, A. D.; and Michelbacher, Abraham E. 1939. "Recommendations for a more discriminating use of insecticides." *Proceedings of the Sixth Pacific Science Congress* 6: 119–123.

Howard, Leland Ossian. 1881. "Report on the parasites of the Coccidae in the collection of this department." *Annual Report of the U.S. Commissioner of Agriculture* 1880: 350–373.

_____. 1897. "A study in insect parasitism: A consideration of the parasites of the white-marked tussock moth, with an account of their habits and interrelations, and with descriptions of new species." *U.S. Division of Entomology Technical Series* 5.

_____. 1898. "Danger of importing insect pests." *U.S. Department of Agriculture Yearbook* 1897: 529–552.

_____. 1906. "An interesting new genus and species of Encyrtidae." *Entomological News* 17: 121–122.

_____. 1907. "The male of *Comperiella*." *Entomological News* 18: 237.

_____. 1910. "Two new aphelinine parasites of scale insects." *Entomological News* 21: 162–163.

_____. 1916. "On the Hawaiian work in introducing beneficial insects." *Journal of Economic Entomology* 9: 172–179.

_____. 1917. "The practical use of the insect enemies of injurious insects." *U.S.*

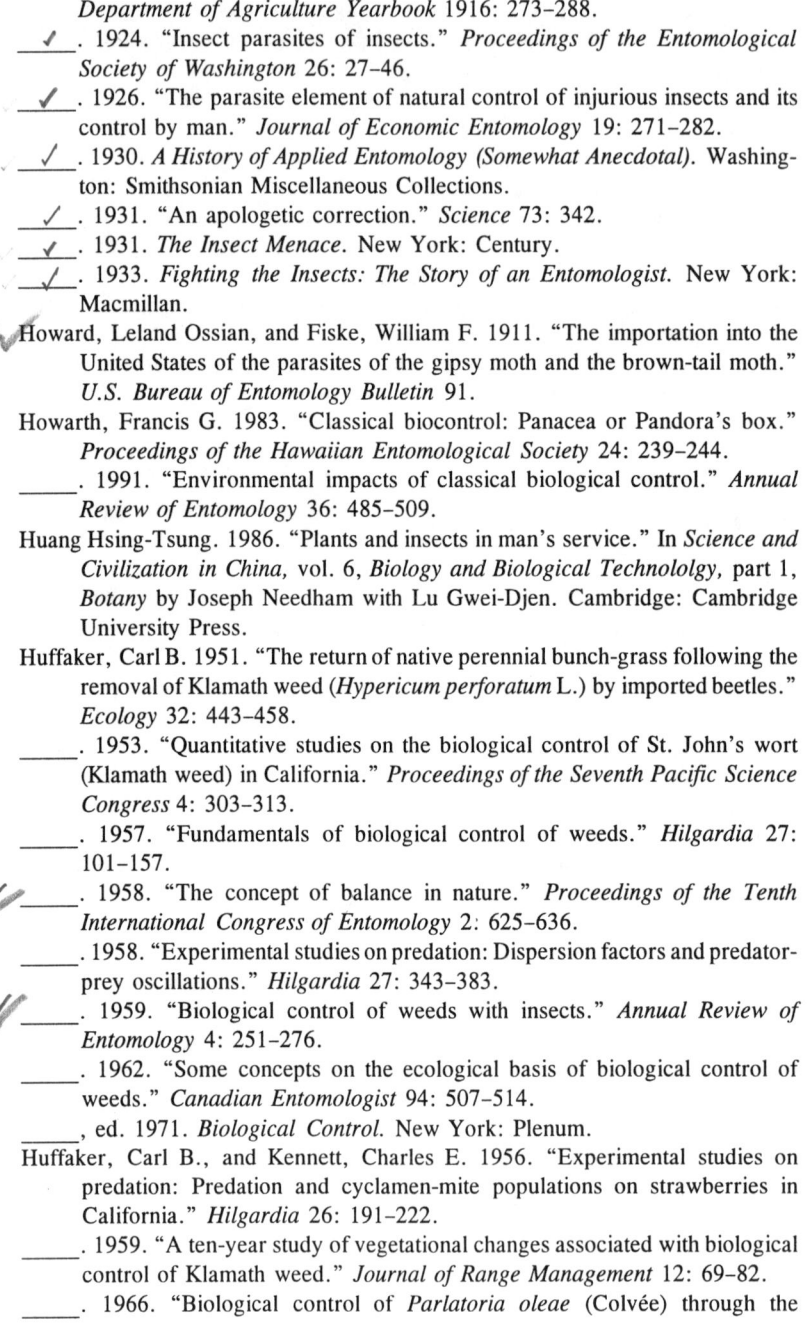

    Department of Agriculture Yearbook 1916: 273–288.

       . 1924. "Insect parasites of insects." *Proceedings of the Entomological Society of Washington* 26: 27–46.

       . 1926. "The parasite element of natural control of injurious insects and its control by man." *Journal of Economic Entomology* 19: 271–282.

       . 1930. *A History of Applied Entomology (Somewhat Anecdotal).* Washington: Smithsonian Miscellaneous Collections.

       . 1931. "An apologetic correction." *Science* 73: 342.

       . 1931. *The Insect Menace.* New York: Century.

       . 1933. *Fighting the Insects: The Story of an Entomologist.* New York: Macmillan.

Howard, Leland Ossian, and Fiske, William F. 1911. "The importation into the United States of the parasites of the gipsy moth and the brown-tail moth." *U.S. Bureau of Entomology Bulletin* 91.

Howarth, Francis G. 1983. "Classical biocontrol: Panacea or Pandora's box." *Proceedings of the Hawaiian Entomological Society* 24: 239–244.

       . 1991. "Environmental impacts of classical biological control." *Annual Review of Entomology* 36: 485–509.

Huang Hsing-Tsung. 1986. "Plants and insects in man's service." In *Science and Civilization in China,* vol. 6, *Biology and Biological Technololgy,* part 1, *Botany* by Joseph Needham with Lu Gwei-Djen. Cambridge: Cambridge University Press.

Huffaker, Carl B. 1951. "The return of native perennial bunch-grass following the removal of Klamath weed (*Hypericum perforatum* L.) by imported beetles." *Ecology* 32: 443–458.

       . 1953. "Quantitative studies on the biological control of St. John's wort (Klamath weed) in California." *Proceedings of the Seventh Pacific Science Congress* 4: 303–313.

       . 1957. "Fundamentals of biological control of weeds." *Hilgardia* 27: 101–157.

       . 1958. "The concept of balance in nature." *Proceedings of the Tenth International Congress of Entomology* 2: 625–636.

       . 1958. "Experimental studies on predation: Dispersion factors and predator-prey oscillations." *Hilgardia* 27: 343–383.

       . 1959. "Biological control of weeds with insects." *Annual Review of Entomology* 4: 251–276.

       . 1962. "Some concepts on the ecological basis of biological control of weeds." *Canadian Entomologist* 94: 507–514.

       , ed. 1971. *Biological Control.* New York: Plenum.

Huffaker, Carl B., and Kennett, Charles E. 1956. "Experimental studies on predation: Predation and cyclamen-mite populations on strawberries in California." *Hilgardia* 26: 191–222.

       . 1959. "A ten-year study of vegetational changes associated with biological control of Klamath weed." *Journal of Range Management* 12: 69–82.

       . 1966. "Biological control of *Parlatoria oleae* (Colvée) through the compensatory action of two introduced parasites." *Hilgardia* 37: 283–335.

Huffaker, Carl B., and Messenger, Powers S., eds. 1976. *Theory and Practice of Biological Control.* New York: Academic Press.

Huxley, Julian. 1942. *Evolution: The Modern Synthesis.* London: G. Allen & Unwin.

———, ed. 1940. *The New Systematics.* Oxford: Clarendon Press.

Imms, Augustus D. 1931. *Recent Advances in Entomology.* Philadelphia: P. Blakiston's Son.

Jacobs, Josephine Kingsbury. 1966. "Sunkist advertising." Ph.D. dissertation, University of California, Los Angeles.

Jaynes, Harold A., and Bynum, E. K. 1941. "Experiments with *Trichogramma minutum* Riley as a control of the sugarcane borer." *U.S. Department of Agriculture Technical Bulletin* 743.

Jelinek, Lawrence J. 1979. *Harvest Empire: A History of California Agriculture.* San Francisco: Boyd & Fraser.

Kennett, Charles, E.; Huffaker, Carl B.; and Finney, Glenn L. 1966. "The role of an autoparasitic aphelinid, *Coccophagoides utilis* Doutt, in the control of *Parlatoria oleae* (Colvée)." *Hilgardia* 37: 255–282.

Kimler, William C. 1986. "Advantage, adaptiveness, and evolutionary ecology." *Journal of the History of Biology* 19: 215–233.

Kingsland, Sharon E. 1985. *Modeling Nature: Episodes in the History of Population Ecology.* Chicago: University of Chicago Press.

Kirby, William, and Spence, William. 1815–1878. *An Introduction to Entomology.* 4 vols. London: Longman.

Kloppenburg, Jack Ralph, Jr. 1988. *First the Seed: The Political Economy of Plant Biotechnology, 1492–2000.* Cambridge: Cambridge: University Press.

Klotz, Esther H.; Lawton, Harry W.; and Hall, Joan H., eds. 1969. *A History of Citrus in the Riverside Area.* Riverside, Calif.: Riverside Museum Press.

Knapp, Joseph G. 1969. *The Rise of American Cooperative Enterprise, 1620–1920.* Danville, Ill.: Interstate.

———. 1973. *The Advance of American Cooperative Enterprise, 1920–1945.* Danville, Ill.: Interstate.

Koebele, Albert. 1890. "Report of a trip to Australia made under the direction of the Entomologist to investigate the natural enemies of the fluted scale." *U.S. Division of Entomology Bulletin* 21.

Köllar, Vincent. 1840. *A Treatise on Insects Injurious to Gardeners, Foresters, and Farmers.* Translated by J. and M. Loudon. London: W. Smith.

Lack, David. 1954. *The Natural Regulation of Animal Numbers.* London: Oxford University Press.

Lawton, Harry W., and Weathers, Lewis G. 1989. "The origins of citrus research in California." In *The Citrus Industry,* vol. 5: *Crop Protection, Postharvest Technology, and Early History of Citrus Research in California,* edited by Walter Reuther, E. Clair Calavan, and Glenn E. Carman. Berkeley: University of California Division of Agriculture and Natural Resources.

Layton, Edwin T., Jr. 1971. "Mirror-image twins: The communities of science and technology in 19th-century America." *Technology and Culture* 12: 562–580.

_____. 1974. "Technology as knowledge." *Technology and Culture* 15: 31–41.

_____. 1987. "Through the looking glass, or news from Lake Mirror Image." *Technology and Culture* 28: 594–607.

Leary, James C.; Fishbein, William I; and Salter, Lawrence C. 1947. *DDT and the Insect Problem.* New York: Macmillan.

Lelong, Byron Martin. 1890. "Beneficial insects." *Annual Report of the California State Board of Horticulture* 1889: 260–288.

_____. 1902. *Culture of the Citrus in California.* Sacramento: A. G. Johnston.

Levenstein, Harvey A. 1988. *Revolution at the Table: The Transformation of the American Diet.* New York: Oxford University Press.

Lloyd, John William. 1919. *Co-operative and Other Organized Methods of Marketing California Horticultural Products.* Urbana: University of Illinois.

Lorbeer, Howard B. 1931. "The organization and operation of a pest control district." *California Citrograph* 16: 451, 484.

Lyle, Clay. 1947. "Achievements and possibilities in pest eradication." *Journal of Economic Entomology* 40: 1–8.

McCook, H. C. 1883. "Ants as beneficial insecticides." *Proceedings of the Academy of Natural Sciences of Philadelphia* 1882: 263–271.

McIntosh, Robert P. 1985. *The Background of Ecology: Concept and Theory.* Cambridge: Cambridge University Press.

McKay, A. W.; Samson, H. W.; Pailthrop, R. R.; Flohr, L. B.; Corbett, L. C.; Hawkins, L. A.; Magness, J. R.; Gould, H. P.; and Beattie, W. R. 1926. "Marketing fruits and vegetables." *U.S. Department of Agriculture Yearbook* 1925: 623–710.

McKenzie, Howard L. 1937. "Morphological differences distinguishing California red scale, yellow scale, and related species (Homoptera—Diaspididae)." *University of California Publications in Entomology* 6: 323–336.

_____. 1964. "Fourth taxonomic study of California mealybugs, with additional species from North America, South America, and Japan (Homoptera: Coccoidea: Pseudococcidae.)" *Hilgardia* 35: 211–272.

Mackerras, Ian M. 1970. "Alexander John Nicholson." *Records of the Australian Academy of Science* 2, no. 1: 66–81.

Mackie, David B. 1944. "One year with the oriental fruit moth." *California Department of Agriculture Bulletin* 33: 4–17.

McLean, Louis A. 1967. "Pesticides and the environment." *Bioscience* 17: 613–617.

McWilliams, Carey. 1946. *Southern California Country: An Island on the Land.* New York: Duell, Sloan, and Pearce.

Mallis, Arnold. 1971. *American Entomologists.* New Brunswick, N.J.: Rutgers University Press.

Manners, Ian P. 1979. "The persistent problem of the boll weevil: Pest control in principle and in practice." *Geographical Review* 69: 25–42.

Marchal, Paul. 1897. "L'équilibre numerique des espèces et ses relations avec les parasites chez les insectes." *Comptes rendus hebdomadaires des séances at memoires de la Société de Biologie* 49: 129–130.

———. 1908. "The utilization of auxiliary entomophagous insects in the struggle against insects injurious to agriculture." *Popular Science Monthly* 72: 352–370, 406–419.

Marcus, Alan I. 1985. *Agricultural Science and the Quest for Legitimacy: Farmers, Agricultural Colleges, and Experiment Stations, 1870–1890.* Ames: Iowa State University Press.

Martinez, Richard. 1986. "University makes neighborhood go." *Riverside Press-Enterprise,* 22 October 1986.

Mayr, Ernst. 1942. *Systematics and the Origin of Species.* Reprint. New York: Dover, 1962.

———. 1982. *The Growth of Biological Thought: Diversity, Evolution, and Inheritance.* Cambridge, Mass: Harvard University Press, Belknap Press.

Mayr, Ernst; Linsley, E. Gorton; and Usinger, Robert L. 1953. *Methods and Principles of Systematic Zoology.* New York: McGraw-Hill.

Mayr, Ernst, and Provine, William B., eds. 1980. *The Evolutionary Synthesis: Perspectives on the Unification of Biology.* Cambridge, Mass.: Harvard University Press.

Meiners, Edwin P. 1959. "Charles Valentine Riley." Mimeographed. Columbia, Mo.: C. V. Riley Entomological Society.

Michelbacher, Abraham E. 1940. "Some entomological observations in California." *Journal of Economic Entomology* 33: 141–143.

———. 1943. "The present status of the alfalfa weevil in California." *California Agricultural Experiment Station Bulletin* 677.

———. 1945. "The importance of ecology in insect control" *Journal of Economic Entomology* 38: 129–130.

Michelbacher, Abraham E., and Bacon, Oscar G. 1952. "Walnut insect and spider-mite control in northern California." *Journal of Economic Entomology* 45: 1020–1027.

Milne, Alec. 1957. "Theories of natural control of insect populations." *Cold Spring Harbor Symposia on Quantitative Biology* 22: 253–271.

———. 1958. "Perfect and imperfect density dependence in population dynamics." *Nature* 182: 1251–1252.

Moffat, Anne Simon. 1991. "Research on biological pest control moves ahead." *Science* 252: 211–212.

Morrill, Austin Winfield. 1931. "A discussion of Smith and Flanders' *Trichogramma* fad query." *Journal of Economic Entomology* 24: 1264–1273.

Morrill, Austin Winfield, and Back, Ernest A. 1912. "Natural control of white flies in Florida." *U.S. Bureau of Entomology Bulletin* 102.

Morse, L. D. 1869. "Notice." *Annual Report of the Missouri State Board of Agriculture* 1868: iii–v.

Mowry, George E. 1951. *The California Progressives.* Berkeley: University of California Press.

Muesebeck, C. F. W. 1942. "Fundamental taxonomic problems in quarantine and nursery inspection." *Journal of Economic Entomology* 35: 753–758.

Nash, Gerald D. 1964. *State Government and Economic Development: A History of Administrative Policies in California, 1849–1933.* Berkeley: Institute of Governmental Studies, University of California.

_____. 1967. "The sugar beet industry and economic growth in the West." *Agricultural History* 41: 27–30.

"Neglected orchards may be destroyed under new law." 1931. *California Citrograph* 16:536.

Nicholson, Alexander John. 1927. "A new theory of mimicry in insects." *Australian Zoologist* 5: 10–104.

_____. 1933. "The balance of animal populations." *Journal of Animal Ecology* 2: 132–178.

_____. 1947. "Fluctuation of animal populations." *Report of the Australia and New Zealand Association for the Advancement of Science* 26: 134–147.

_____. 1954. "An outline of the dynamics of animal populations." *Australian Journal of Zoology* 2: 9–65.

"No excuse for this." 1942. *California Citrograph* 27: 65.

Nye, Ronald L. 1983. "Federal vs. state agricultural research policy: The case of California's Tulare experiment station." *Agricultural History* 57: 436–449.

O'Connor, B. A. 1953. "Biological control of insects and plants in Fiji." *Proceedings of the Seventh Pacific Science Congress* 4: 278–293.

Olin, Spencer C., Jr. 1968. *California's Prodigal Sons: Hiram Johnson and the Progressives, 1911–1917.* Berkeley: University of California Press.

Olmstead, Alan L., and Rhode, Paul. 1988. "An overview of California agricultural mechanization, 1870–1930." *Agricultural History* 62, no. 3: 86–112.

Ordish, George. 1976. *The Constant Pest: A Short History of Pests and their Control.* London: Peter Davies.

Orsi, Richard J. 1975. "*The Octopus* revisited: The Southern Pacific and agricultural modernization in California." *California Historical Quarterly* 54: 197–220.

Osborn, Herbert. 1937–46. *Fragments of Entomological History, Including some Personal Recollections of Men and Events.* 2 vols. Columbus, Ohio.

Osborne, Thomas J. 1981. *"Empire can Wait": American Opposition to Hawaiian Annexation, 1893–1898.* Kent, Ohio: Kent State University Press.

Overfield, Richard A. 1986. "The agricultural experiment station and Americanization: The Hawaiian experience, 1900–1910." *Agricultural History* 60, no. 2: 256–266.

Packard, Alpheus Spring, Jr. 1895. "Charles Valentine Riley." *Science,* n.s. 2: 745–751.

Palladino, Paolo S. A. 1989. "Entomology and ecology: The ecology of entomology. The insecticide crisis and entomological research in the United States in the 1960s and 1970s: Political, institutional, and conceptual dimensions." Ph.D. dissertation, University of Minnesota.

_____. 1990. "Ecological theory and pest control practice: A study of the institutional and conceptual dimensions of a scientific debate." *Social Studies of Science* 20: 255–281.

Paterson, Alan M. 1975. "Oranges, soot, and science: The development of frost protection in California." *Technology and Culture* 16: 360–376.

Patterson, Tom. 1964. *Landmarks of Riverside and the Stories Behind Them.* Riverside, Calif.: Press-Enterprise Co.

_____. 1969. "Chemical, biological advocates join forces at UCR." *Riverside Daily Enterprise,* 31 July 1969.

_____. 1971. *A Colony for California: Riverside's First Hundred Years.* Riverside, Calif.: Press-Enterprise Co.

Pemberton, Cyril E. 1948. "History of the Entomology Department Experiment Station, HSPA, 1904–45." *Hawaiian Planters Record* 52, no. 1: 53–90.

_____. 1953. "The biological control of insects in Hawaii." *Proceedings of the Seventh Pacific Science Congress* 4: 220–223.

_____. 1964. "Highlights in the history of entomology in Hawaii, 1778–1963." *Pacific Insects* 6: 689–729.

Pemberton, Cyril E., and Willard, Harold F. 1917. "Parasitism of the larvae of the Mediterranean fruit fly in Hawaii during 1916." *Report of the Board of Commissioners of Agriculture and Forestry of the Territory of Hawaii 1915–16:* 111–118.

_____. 1918. "Interrelations of fruit-fly parasites in Hawaii." *Journal of Agricultural Research* 12: 285–295.

Perkins, John H. 1978. "Edward Fred Knipling's sterile-male technique for control of the screwworm fly." *Environmental Review* 1, no. 5: 19–37.

_____. 1981. *Insects, Experts, and the Insecticide Crisis: The Quest for New Pest Management Strategies.* New York: Plenum.

_____. 1983. "Insects, food, and hunger: The paradox of plenty for U.S. entomology, 1920–1970." *Environmental Review* 7: 71–96.

Perkins, R. C. L., and Swezey, Otto H. 1924. "The introduction into Hawaii of insects that attack lantana." *Hawaiian Sugar Planters' Association Experiment Station, Entomological Series, Bulletin* 16.

Peterson, Alvah. 1930. "How many species of *Trichogramma* occur in North America?" *Journal of the New York Entomological Society* 38: 1–8.

Pickett, A. D. 1948. "The philosophy of orchard insect control." *Annual Report of the Entomological Society of Ontario* 79: 37–41.

_____. 1949. "A critique on insect chemical control methods." *Canadian Entomologist* 81: 67–76.

Pickett, A. D., and Patterson, N. A. 1953. "The influence of spray programs on the fauna of apple orchards in Nova Scotia. IV. A review." *Canadian Entomologist* 85: 472–478.

Pickett, A. D.; Patterson, N. A.; Stultz, H. T.; and Lord, F. T. 1946. "The influence of spray programs on the fauna of apple orchards in Nova Scotia. I. An appraisal of the problem and a method of approach." *Scientific Agriculture* 26: 590–600.

Pickett, A. D.; Putman, William L.; and LeRoux, E. J. 1958. "Progress in harmonizing biological and chemical control of orchard pests in eastern Canada." *Proceedings of the Tenth International Congress of Entomology* 3: 169–174.

"The picture changes." 1942. *California Citrograph* 27: 65.

Pimentel, David, and Perkins, John H. 1980. *Pest Control: Cultural and Environmental Aspects*. Boulder, Colo.: Westview Press.

Pisani, Donald J. 1984. *From the Family Farm to Agribusiness: The Irrigation Crusade in California and the West, 1850–1931*. Berkeley: University of California Press.

"Ponder this." 1924. *California Citrograph* 10: 41.

Powell, G. Harold. 1908. "The decay of oranges while in transit from California." *U.S. Bureau of Plant Industry Bulletin* 123.

Prizer, J. A. 1925. "Pest control and its relation to marketing of citrus." *California Citrograph* 10: 197, 232.

"Prof. C. V. Riley, M.A., Ph.D." 1895. *Entomological News* 6: 241–243.

"Putting bugs to work." 1948. *California Citrograph* 33: 377.

Quayle, Henry J. 1908. "A statistical study of brown scale parasitism." *Science* 27: 788–789.

_____. 1910. "*Scutellista cyanea* Motsch." *Journal of Economic Entomology* 3: 446–451.

_____. 1911. "Citrus fruit insects." *California Agricultural Experiment Station Bulletin* 214: 443–512.

_____. 1911. "Scale insect parasitism in California." *Journal of Economic Entomology* 4: 510–515.

_____. 1938. "The development of resistance to hydrocyanic acid in certain scale insects." *Hilgardia* 11: 183–210.

_____. 1938. *Insects of Citrus and Other Subtropical Fruits*. Ithaca, N.Y.: Comstock.

Rector, Thomas A. 1950. "Frozen concentrated orange juice—its research background." *Refrigerating Engineering* 58: 349–353.

Reed, John Henry. 1904. "California Farmers' Clubs." *Pacific Rural Press* 67: 278.

Riley, Charles Valentine. 1870. "In memoriam [Benjamin D. Walsh]." *American Entomologist* 2: 65–68.

_____. 1870. *Second Annual Report on the Noxious, Beneficial, and Other Insects of the State of Missouri*. Jefferson City: Horace Wilcox.

_____. 1871. *Third Annual Report on the Noxious, Beneficial, and Other Insects of the State of Missouri*. Jefferson City: Horace Wilcox.

_____. 1872. *Fourth Annual Report on the Noxious, Beneficial, and Other Insects of the State of Missouri*. Jefferson City: Regan & Edwards.

_____. 1873. *Fifth Annual Report on the Noxious, Beneficial, and Other Insects of the State of Missouri*. Jefferson City: Regan & Carter.

_____. 1874. *Sixth Annual Report on the Noxious, Beneficial, and Other Insects of the State of Missouri*. Jefferson City: Regan & Carter.

_____. 1879. "Report of the entomologist." *Annual Report of the U.S. Commis-*

sioner of Agriculture 1878: 207–257.

___√__. 1880. "Large white scale on acacias, etc." *American Entomologist* 3: 20.

___√__. 1884. "Report of the entomologist." *Annual Report of the U.S. Commissioner of Agriculture* 1883: 99–180.

___√__. 1885. "Report of the entomologist." *Annual Report of the U.S. Commissioner of Agriculture* 1884: 285–418.

___√__. 1887. "Address of Professor C. V. Riley [at Seventh State Fruit Growers Convention, Riverside.]" *Biennial Report of the California State Board of Horticulture* 1885–86: 450–462; with discussion, 462–471.

___√__. 1887. "The scale insects of the orange in California, and particularly the *Icerya* or fluted scale, *alias* white scale, *alias* cottony-cushion scale, etc." *Pacific Rural Press* 33: 361–364.

___√__. 1888. "On the original habitat of *Icerya purchasi.*" *Pacific Rural Press* 35: 425.

___√__. 1888. "Report of the entomologist." *Annual Report of the U.S. Commissioner of Agriculture* 1887: 48–179.

___√__. 1892. "Parasitism in insects." *Proceedings of the Entomological Society of Washington* 2: 397–431.

_____. 1893. "Parasitic and predaceous insects in applied entomology." *Insect Life* 6: 130–141.

Rosenberg, Charles E. 1976. *No Other Gods: On Science and American Social Thought.* Baltimore: Johns Hopkins University Press.

Rossiter, Margaret W. 1979. "The organization of the agricultural sciences." In *The Organization of Knowledge in Modern America, 1860–1920,* edited by Alexandra Oleson and John Voss, pp. 211–248. Baltimore: Johns Hopkins University Press.

_____. 1982. *Women Scientists in America: Struggles and Strategies to 1940.* Baltimore: Johns Hopkins University Press.

Rothstein, Morton. 1975. "West coast farmers and the tyranny of distance: Agriculture on the fringes of the world market." *Agricultural History* 49: 272–280.

Sailer, Reece I. 1972. "A look at the USDA's biological control of insect pests: 1888 to present." *Agricultural Science Review* 10, no. 4: 15–27.

Sailer, Reece I.; Brown, R. E.; Munir, B.; and Nickerson, J. C. E. 1984. "Dissemination of the citrus whitefly (Homoptera: Aleyrodidae) parasite *Encarsia lahorensis* (Howard) (Hymenoptera: Aphelinidae) and its effectiveness as a control agent in Florida." *Bulletin of the Entomological Society of America* 30: 36–39.

Salt, George. 1934. "Experimental studies in insect parasitism. II. Superparasitism." *Proceedings of the Royal Society of London, Series B* 114: 455–476.

_____. 1937. "The sense used by *Trichogramma* to distinguish between parasitized and unparasitized hosts." *Proceedings of the Royal Society of London, Series B* 122: 57–75.

Sampson, Arthur W. 1919. "Plant succession in relation to range management." *U.S. Department of Agriculture Bulletin* 791.

Sampson, Arthur W., and Malmsten, Harry E. 1935. "Stock-poisoning plants of California." *California Agricultural Experiment Station Bulletin* 593.

Sampson, Arthur W., and Parker, Kenneth W. 1930. "St. Johnswort on range lands of California." *California Agricultural Experiment Station Bulletin* 503.

Sapp, Jan. 1987. *Beyond the Gene: Cytoplasmic Inheritance and the Struggle for Authority in Genetics.* New York: Oxford University Press.

Sawyer, Richard C. 1990. "Monopolizing the insect trade: Biological control in the USDA, 1888–1951." *Agricultural History* 64, no. 2: 271–285.

Schedvin, Carl Boris. 1987. "Environment, economy, and Australian biology, 1890–1939." In *Scientific Colonialism: A Cross-cultural Comparison,* edited by Nathan Reingold and Marc Rothenberg, pp. 101–126. Washington: Smithsonian Institution Press.

_____. 1987. *Shaping Science and Industry: A History of Australia's Council for Scientific and Industrial Research, 1926–49.* Sydney: Allen & Unwin.

Scott, Roy V. 1970. *The Reluctant Farmer: The Rise of Agricultural Extension to 1914.* Urbana: University of Illinois Press.

"Search for scale parasites." 1931. *Citrus News* 7: 3, 18.

Seftel, Howard. 1985. "Government regulation and the rise of the California fruit industry: The entrepreneurial attack on fruit pests, 1880–1920." *Business History Review* 59: 369–402.

Severin, Henry H. P. 1912. "The introduction, methods of control, spread, and migration of the Mediterranean fruit fly in the Hawaiian Islands." *California State Commission of Horticulture Bulletin* 1: 558–565.

_____. 1924. "Natural enemies of beet leafhopper (*Eutettix tenella* Baker)." *Journal of Economic Entomology* 17: 369–377.

Shelford, Victor E. 1937. *Animal Communities in Temperate America as Illustrated in the Chicago Region,* 2nd ed. Chicago: University of Chicago Press.

Shinn, Charles Howard. 1891. "Social changes in California." *Popular Science Monthly* 38: 794–803.

Shortland, Michael, ed. 1993. *Science and Nature: Essays in the History of the Environmental Sciences.* [Faringdon, Engl.]: British Society for the History of Science.

Smith, Harry S. 1912. "The chalcidoid genus *Perilampus* and its relation to the problem of parasite introduction." *U.S. Bureau of Entomology Technical Series* 19: 33–69.

_____. 1914. "Mealy bug parasites in the Far East." *California State Commission of Horticulture Bulletin* 3: 26–29.

_____. 1914. "The season's work with *Hippodamia convergens.*" *California State Commission of Horticulture Bulletin* 3: 77–78.

_____. 1915. "Report of the State Insectary." *California State Commission of Horticulture Bulletin* 4: 542–543.

_____. 1915. "Some misconceptions." *California State Commission of Horticulture Bulletin* 4: 269–270.

_____. 1915. "The use of fungous diseases against the black scale." *California State Commission of Horticulture Bulletin* 4: 109–111.

_____. 1916. "An attempt to redefine the host relationships exhibited by entomophagous insects." *Journal of Economic Entomology* 9: 477–486.

_____. 1916. "Beet leaf-hopper parasites." *California State Commission of Horticulture Bulletin* 5: 299.

_____. 1916. "A sublaboratory of the Insectary in the south." *California State Commission of Horticulture Bulletin* 5: 307.

_____. 1917. "The alfalfa weevil." *California State Commission of Horticulture Bulletin* 6: 295–297.

_____. 1917. "An Australian expedition from the Insectary." *California State Commission of Horticulture Bulletin* 6: 33.

_____. 1917. "Economy in insect control through the use of parasites." *California State Commission of Horticulture Bulletin* 6: 362.

_____. 1917. "The habit of leaf oviposition among the parasitic Hymenoptera." *Psyche* 24: 63–68.

_____. 1917. "On the life-history and successful introduction into the United States of the Sicilian mealy-bug parasite." *Journal of Economic Entomology* 10: 262–268.

_____. 1918. "Everett Jay Vosler." *California State Commission of Horticulture Bulletin* 7: inside title page.

_____. 1919. "Biennial report of the Insectary Division, State Commission of Horticulture, 1917–18." *California State Commission of Horticulture Bulletin* 8: 44–51.

_____. 1919. "On some phases of insect control by the biological method." *Journal of Economic Entomology* 12: 288–292.

_____. 1921. "Biological control of the black scale in California." *California Department of Agriculture Bulletin* 10: 127–137.

_____. 1921. "Report of the Bureau of Pest Control." *California Department of Agriculture Bulletin* 10: 570–597.

_____. 1923. "Biological control work." *California Department of Agriculture Bulletin* 12: 334–342.

_____. 1923. "What may we expect from biological control?" *Journal of Economic Entomology* 16: 506–511.

_____. 1925. "The commercial development of biological control in California." *Journal of Economic Entomology* 18: 147–152.

_____. 1926. "The fundamental importance of life-history data in biological control work." *Journal of Economic Entomology* 19: 708–714.

_____. 1926. "Status of biological control work of citrus pests." *California Citrograph* 11: 414, 426–431.

_____. 1929. "Multiple parasitism: Its relation to the biological control of insect pests." *Bulletin of Entomological Research* 20: 141–149.

_____. 1929. "On some phases of preventive entomology." *Scientific Monthly* 29: 177–184.

_____. 1929. "Present status of biological control of mealybugs." *California*

Citrograph 14: 428–429.

_____. 1929. "The utilization of entomophagous insects in the control of citrus pests." *Transactions of the Fourth International Congress of Entomology* 2: 191–198.

_____. 1930. "Plant quarantines, their aims, and their biological and economic justification." *California Department of Agriculture Bulletin* 19: 203–207.

_____. 1931. "Is citrophilus mealybug under permanent control?" *California Citrograph* 16: 429, 448.

_____. 1933. "The influence of civilization on the insect fauna by purposeful introductions." *Annals of the Entomological Society of America* 26: 518–528.

_____. 1934. "University of California to continue red scale parasite search." *California Citrograph* 19: 263, 280–282.

_____. 1935. "The role of biotic factors in the determination of population densities." *Journal of Economic Entomology* 28: 873–898.

_____. 1937. Review of *The Biological Control of Insects*, by Harvey L. Sweetman. *Journal of Economic Entomology* 30: 218–220.

_____. 1939. "Insect populations in relation to biological control." *Ecological Monographs* 9: 311–320.

_____. 1940. "Economic entomology and the national welfare." *Journal of Economic Entomology* 33: 587–588.

_____. 1941. "Racial segregation of insect populations and its significance in applied entomology." *Journal of Economic Entomology* 34: 1–13.

_____. 1941. "Status of biological control of scale pests." *California Citrograph* 26: 58, 76–77.

_____. 1942. "A race of *Comperiella bifasciata* successfully parasitizes California red scale." *Journal of Economic Entomology* 35: 809–812.

_____. 1944. "Julian Huxley on evolution." *Ecology* 25: 477–479.

_____. 1945. "Cost of producing *Macrocentrus* by the potato-tuber-worm method." *Journal of Economic Entomology* 38: 316–319.

_____. 1946. "The biological control program in relation to California agriculture." *California Citrograph* 31: 414–415, 452–453.

_____. 1948. "Biological control of insect pests." In *The Citrus Industry*, vol. 2: *Production of the Crop*, edited by Leon D. Batchelor and Herbert J. Webber. Berkeley: University of California Press.

_____. 1955. "Ecological aspects of insect population dynamics." Paper read at Entomological Society of America meeting, Cincinnati.

Smith, Harry S., and Armitage, Horace Morton. 1920. "Biological control of mealybugs in California." *California Department of Agriculture Bulletin* 9: 104–158.

_____. 1931. "The biological control of mealybugs attacking citrus." *California Agricultural Experiment Station Bulletin* 509.

Smith, Harry S., and Basinger, Almon J. 1947. "History of biological control." *California Cultivator* 94: 720, 729, 754–755.

Smith, Harry S., and Compere, Harold. 1920. "The life-history and successful introduction into California of the black scale parasite, *Aphycus lounsburyi* How." *California Department of Agriculture Bulletin* 9: 310–320.

_____. 1928. "The introduction of new insect enemies of the citrophilus mealybug from Australia." *Journal of Economic Entomology* 21: 664–669.

_____. 1931. "Experiment station new insectary and biological control program." *California Citrograph* 16: 236–237.

_____. 1931. "An imported parasite attacks the yellow scale." *California Citrograph* 16: 328.

_____. 1931. "Introduced parasites successfully control the citrophilus mealybug." *Journal of Economic Entomology* 24: 942–945.

Smith, Harry S., and DeBach, Paul. 1942. "The measurement of the effect of entomophagous insects on population densities of their hosts." *Journal of Economic Entomology* 35: 845–849.

Smith, Harry S.; Essig, Edward O.; Quayle, Henry J.; Fawcett, Howard S.; Smith, Ralph E.; Peterson, George M.; and Tolley, Howard R. 1933. "The efficacy and economic effects of plant quarantines in California." *California Agricultural Experiment Station Bulletin* 553.

Smith, Harry S., and Flanders, Stanley E. 1931. "Is *Trichogramma* becoming a fad?" *Journal of Economic Entomology* 24: 666–672.

_____. 1950. "The search for natural enemies of citrus pests." *California Citrograph* 35: 362, 376–378.

Smith, Harry S., and Vosler, Everett J. 1914. "*Calliephialtes* in California." *California State Commission of Horticulture Bulletin* 3: 195–211.

Smith, Ray F. 1974. "The origins of integrated control in California, an account of the contributions of Charles W. Woodworth." *Pan-Pacific Entomologist* 50: 426–440.

Smith, Ray F., and Allen, William W. 1954. "Insect control and the balance of nature." *Scientific American* 190, no. 6: 38–42.

Smith, Ray F., and Hagen, Kenneth S. 1959. "Integrated control programs in the future of biological control." *Journal of Economic Entomology* 52: 1106–1108.

Smith, Ray F.; Mittler, Thomas E.;, and Smith, Carroll N., eds. 1973. *History of Entomology*. Palo Alto, Calif.: Annual Reviews.

Solomon, M. E. 1949. "The natural control of animal populations." *Journal of Animal Ecology* 18: 1–35.

_____. 1957. "Dynamics of insect populations." *Annual Review of Entomology* 2: 121–142.

_____. 1958. "Meaning of density-dependence and related terms in population dynamics." *Nature* 181: 1778–1780.

_____. 1958. *Nature* 182: 1252.

Sorensen, Willis Conner. 1984. "Brethren of the net: American entomology, 1840–1880." Ph.D. dissertation, University of California, Davis.

_____. 1988. "The rise of government sponsored applied entomology, 1840–1870." *Agricultural History* 62, no. 2: 98–115.

Spangler, Raymond L. 1946. "Standardization and inspection of fresh fruits and vegetables." *U.S. Department of Agriculture Miscellaneous Publication* 604.

Stadtman, Verne A. 1970. *The University of California, 1868-1968*. New York: McGraw-Hill.

_____, ed. 1967. *The Centennial Record of the University of California*. Berkeley: University of California Printing Department.

Starr, Kevin. 1973. *Americans and the California Dream, 1850-1915*. New York: Oxford University Press.

_____. 1985. *Inventing the Dream: California through the Progressive Era*. New York: Oxford University Press.

Staudenmaier, John M., S.J. 1985. *Technology's Storytellers: Reweaving the Human Fabric*. Cambridge, Mass.: Society for the History of Technology and MIT Press.

Steinhaus, Edward A. 1951. "Possible use of *Bacillus thuringiensis* Berliner as an aid in the biological control of the alfalfa caterpillar." *Hilgardia* 20: 359-381.

_____. 1956. "Potentialities for microbial control of insects." *Journal of Agricultural and Food Chemistry* 4: 676-680.

_____. 1975. *Disease in a Minor Chord: Being a Semihistorical and Semibiographical Account of a Period in Science When One Could Be Happily Yet Seriously Concerned with the Diseases of Lowly Animals Without Backbones, Especially the Insects*. Columbus: Ohio State University Press.

Stern, Vernon M.; Schlinger, Evert I.; and Bowen, William R. 1965. "Dispersal studies of *Trichogramma semifumatum* (Hymenoptera: Trichogrammatidae) tagged with radioactive phosphorus." *Annals of the Entomological Society of America* 58: 234-240.

Stern, Vernon M.; Smith, Ray F.; van den Bosch, Robert; and Hagen, Kenneth S. 1959. "The integrated control concept." *Hilgardia* 29: 81-101.

Strickland, E. H. 1945. "Could the widespread use of DDT be a disaster?" *Entomological News* 56: 85-88.

"Supervised pest control program of field crop growers." 1949. *California Citrograph* 24: 314-315.

"Supreme court sustains fruit standardization law." 1931. *California Citrograph* 16: 367, 370.

Swazey, Judith P., and Reeds, Karen. 1978. *Today's Medicine, Tomorrow's Science: Essays on Paths of Discovery in the Biomedical Sciences*. Washington: U.S. Department of Health, Education, and Welfare, Public Health Service, National Institutes of Health.

Sweetman, Harvey L. 1936. *The Biological Control of Insects*. Ithaca, N.Y.: Comstock.

Swenk, Myron M. 1937. "In memoriam—Lawrence Bruner." *Nebraska Bird Review* 5: 35-48.

Taylor, Thomas H. C. 1935. "The campaign against *Aspidiotus destructor*, Sign., in Fiji." *Bulletin of Entomological Research* 26: 1-102.

_____. 1937. *The Biological Control of an Insect in Fiji: An Account of the Coconut Leaf-mining Beetle and its Parasite Complex*. London: Imperial

Institute of Entomology.

_____. 1955. "Biological control of insect pests." *Annals of Applied Biology* 42: 190–196.

Teague, Charles C. 1921. "A tribute to the memory of Dr. A. J. Cook." *California Citrograph* 6: 282, 293.

_____. 1925. "Quality and quantity and their relations to successful marketing of citrus fruits." *California Citrograph* 10: 194, 214–215

_____. 1939. *Ten Talks on Citrus Marketing.* Los Angeles.

_____. 1944. *Fifty Years a Rancher: The Recollections of Half a Century Devoted to the Citrus and Walnut Industries of California and to Furthering the Co-operative Movement in Agriculture.* Los Angeles: Ward Ritchie Press.

Thompson, J. M. 1938. "The orange industry—an economic study." *California Agricultural Experiment Station Bulletin* 622.

Thompson, William Robin. 1923. "A criticism of the 'sequence' theory of parasitic control." *Annals of the Entomological Society of America* 16: 115–128.

_____. 1923. "La théorie mathématique de l'action des parasites entomophages." *Revue Générale des Sciences Pures et Appliquées* 34: 202–210.

_____. 1928. "A contribution to the study of biological control and parasitic introduction in continental areas." *Parasitology* 20: 90–112.

_____. 1929. "On natural control." *Parasitology* 21: 269–281.

_____. 1929. "On the relative value of parasites and predators in the biological control of insect pests." *Bulletin of Entomological Research* 19: 343–350.

_____. 1930. *The Biological Control of Insect and Plant Pests: A Report on the Organization and Progress of the Work of Farnham House Laboratory.* London: King's Printer.

_____. 1930. "The principles of biological control." *Annals of Applied Biology* 17: 306–338.

_____. 1937. "Research on biological control." *Nature* 139: 552.

_____. 1939. "Biological control and the theories of the interactions of populations." *Parasitology* 31: 299–388.

_____. 1948. "Can economic entomology be an exact science?" *Canadian Entomologist* 80: 49–55.

_____. 1956. "The fundamental theory of natural and biological control." *Annual Review of Entomology* 1: 379–402.

Thorpe, William H. 1930. "Biological races in insects and allied groups." *Biological Reviews* 5: 177–212.

_____. 1930. "The biology, post-embryonic development, and economic importance of *Cryptochaetum iceryae* (Diptera, Agromyzidae) parasitic on *Icerya purchasi* (Coccidae, Monophlebini)." *Proceedings of the Royal Zoological Society of London* 1930: 929–971.

Thorpe, William H., and Jones, F. G. W. 1937. "Olfactory conditioning in a parasitic insect and its relation to the problem of host selection." *Proceedings of the Royal Society of London, Series B* 124: 56–81.

Timberlake, Philip H. 1927. "Biological control of insect pests in the Hawaiian

Islands." *Proceedings of the Hawaiian Entomological Society* 6: 529–556.

Tobey, Ronald C. 1981. *Saving the Prairies: The Life Cycle of the Founding School of American Plant Ecology, 1895–1955.* Berkeley: University of California Press.

Turnbull, Albert L., and Chant, Donald A. 1961. "The practice and theory of biological control of insects in Canada." *Canadian Journal of Zoology* 39: 697–753.

Ullyett, Gerald C. 1948. "Insecticide programs and biological control in South Africa." *Journal of Economic Entomology* 41: 337–339.

_____. 1951. "Insects, man, and the environment." *Journal of Economic Entomology* 44: 459–464.

University of California. 1954. *Free Enterprise and University Research: A Record of Public Service by Business, Industry, Private Associations, and Foundations, through Support of the Research Activities of the University of California.* Berkeley.

Usinger, Robert L. 1964. "The role of Linnaeus in the advancement of entomology." *Annual Review of Entomology* 9: 1–16.

Uvarov, Boris P. 1921. "A revision of the genus *Locusta*, L. (= *Pachytylus*, Fieb.), with a new theory as to the periodicity and migrations of locusts." *Bulletin of Entomological Research* 12: 135–163.

_____. 1931. "Insects and climate." *Transactions of the Entomological Society of London* 79: 1–247.

Van den Bosch, Robert. 1978. *The Pesticide Conspiracy.* Garden City, N.Y.: Doubleday.

_____. 1989. *The Pesticide Conspiracy.* Reprinted, with new foreword. Berkeley: University of California Press.

Van den Bosch, Robert; Messenger, Powers S.; and Gutierrez, Andrew P. 1982. *An Introduction to Biological Control.* New York: Plenum.

Varley, George C. 1957. "Ecology as an experimental science." *Journal of Animal Ecology* 26: 251–261.

Volterra, Vito. 1931. "Variations and fluctuations in the number of individuals in animal species living together," translated by Mary Evelyn Wells. In *Animal Ecology with Especial Reference to Insects,* by Royal N. Chapman, pp. 409–448. New York: McGraw-Hill.

Vosler, Everett J. 1919. "Some work of the Insectary Division in connection with the attempted introduction of natural enemies of the beet leafhopper." *California State Commission of Horticulture Bulletin* 8: 231–239.

Walsh, Benjamin D. 1864. "On phytophagic varieties and phytophagic species." *Proceedings of the Entomological Society of Philadelphia* 3: 403–430.

_____. 1866. "Imported insects—the gooseberry sawfly." *Practical Entomologist* 1: 117–125.

_____. 1867. "Importing European parasites." *Practical Entomologist* 2: 54–55.

Ward, Ronald W., and Kilmer, Richard L. 1989. *The Citrus Industry: A Domestic and International Economic Perspective.* Ames: Iowa State University Press.

Warner, John Harley. 1980. "Physiological theory and therapeutic explanation in the 1860s: The British debate on the medical use of alcohol." *Bulletin of the History of Medicine* 54: 235–257.

Wasser, C. H. 1977. "Early development of technical range management, ca. 1895–1945." *Agricultural History* 51: 63–77.

Watson, Joseph R. 1926. "Citrus insects and their control." *Florida Agricultural Experiment Station Bulletin* 183: 289–423.

Webber, Herbert J.; Reuther, Walter; and Lawton, Harry W. 1967. "History and development of the citrus industry." In *The Citrus Industry*, vol. 1: *History, World Distribution, Botany, and Varieties*, edited by Walter Reuther, Herbert J. Webber, and Leon D. Batchelor, pp. 1–39. Berkeley: University of California Division of Agricultural Sciences.

Weeks, Jerry Woods. 1977. "Florida gold: The emergence of the Florida citrus industry, 1865–1895." Ph.D. dissertation, University of North Carolina, Chapel Hill.

Weinland, Henry A. 1957. *Now the Harvest: Memories of a County Agricultural Agent*. New York: Exposition Press.

Wellman, Harry R. 1936. "Some economic aspects of regulation shipments of California oranges." *California Agricultural Experiment Station Circular* 338.

West's Annotated California Codes: Food and Agricultural Code. 1986. St. Paul: West Publishing Co.

Westwood, John Obadiah. 1839–40. *An Introduction to the Modern Classification of Insects, Founded on the Natural Habits and Corresponding Organization of the Different Families*. 2 vols. London.

Wheeler, William Morton. 1911. "Insect parasitism and its peculiarities." *Popular Science Monthly* 79: 431–449.

_____. 1928. *The Social Insects, their Origin and Evolution*. New York: Harcourt, Brace.

Whitten, Jamie L. 1966. *That We May Live*. Princeton, N.J.: Van Nostrand.

Whorton, James C. 1974. *Before Silent Spring: Pesticides and Public Health in Pre-DDT America*. Princeton: Princeton University Press.

[Wickson, Edward J.] 1879. "Parasites of scale insects." *Pacific Rural Press,* 27 September 1879, p. 200.

Wilkes, A. 1947. "The effects of selective breeding on the laboratory propagation of insect parasites." *Proceedings of the Royal Society of London, Series B* 134: 227–245.

Wilson, Frank. 1943. "The entomological control of St. John's Wort (*Hypericum perforatum* L.), with particular reference to the insect enemies of the weed in southern France." *Australian Council of Scientific and Industrial Research Bulletin* 169.

_____. 1953. "Some aspects of the control of weeds by insects." *Proceedings of the Seventh Pacific Science Congress* 3: 294–299.

Wiser, Vivian. 1974. "Protecting American agriculture: Inspection and quarantine of imported plants and animals." *U.S. Department of Agriculture Agricul-*

tural Economics Report 266.

Woglum, Russell S. 1913. "Report of a trip to India and the Orient in search of the natural enemies of the citrus white fly." *U.S. Bureau of Entomology Bulletin* 120.

Woodworth, Charles W. 1899. "Orchard fumigation." *California Agricultural Experiment Station Bulletin* 122.

_____. 1903. "Insect pests in California." *Journal of the Department of Agriculture of Western Australia* 8: 564–567.

_____. 1908. "The theory of the parasitic control of insect pests." *Science* 28: 227–230.

Worster, Donald. 1985. *Nature's Economy: A History of Ecological Ideas.* Cambridge: Cambridge University Press.

Index